Encyclopaedic Dictionary of

BIOLOGICAL SCIENCES

ENCYCLOPAEDIC DICTIONARY OF BIOLOGICAL SCIENCES

Vol. 1

K. NARANG
R. RANI
M. PRAKASH

ANMOL PUBLICATIONS PVT LTD
New Delhi-110002

ANMOL PUBLICATIONS PVT LTD
4374/4B, Ansari Road
Daryaganj, New Delhi-110 002

Encyclopaedic Dictionary of Biological Sciences

First Edition, 1991
ISBN 81-7041-462-8 (Set)
ISBN 81-7041-463-6 (Vol.1)
Reprint 1994
Reprint ... 1998

PRINTED IN INDIA
Published by J.L. Kumar for Anmol Publications Pvt. Ltd. New Delhi and Printed at Mehra Offset Press.

Preface

The *Encyclopaedic Dictionary of Biological Sciences* (set of six volumes) has been carefully compiled and edited to meet the long felt needs of increasingly large number of, those who have to deal with the different aspects of biological sciences in colleges, universities, and research institutes. These volumes provide a speedy means of reference to easily interpreted definitions of the terms selected from botany, zoology, environmental science, ecology, genetics, biochemistry, forestry, agriculture geomorphology, cell biology, plant pathology, anatomy, evolution and ecology.

This set of dictionary is more or less encyclopaedic in its contents and style of representation which is evident from the comprehensiveness of its coverage and contextual relevance of explanations with illustrations and chemical formulae wherever necessary.

The entries in this dictionary have been written in a clear and explanatory style to provide both straight forward definitions and invaluable background information. The presentation throughout has been aimed at sustaining the interest of readers while enriching their vocabulary and comprehension of technical terms and expressions.

Throughout the process of compilation and editing these volumes our sincere effort was to make the dictionary as a ready reference for undergraduate and post graduate biological, agricultural, environmental, medical, marine and biochemistry students and also for those appearing for competitive examinations, research scholars, teachers and authors.

The editors express their thanks to their friends and colleagues whose continuous inspiration have initiated them to bring out these volumes.

The Editors

A

A—. Not, not having, the opposite of : a carpellous, apetallous, aseptate, asynchronous, atypical.

A. A class of flowers having freely exposed nectar.

Aardvark. A strange African mammal with no close relatives. It has a bulky grey body about one metre in length, a long narrow head and large ears. The aardvark feeds mainly on termites, which is digs out of the ground with its powerful claws and laps up with a long sticky tongue.

Aardwolf. An African member of the hyena family, somewhat larger than a fox. The coat is yellowish grey with black stripes. It is a nocturnal animal and feeds mainly on termites, which it picks up with its long tongue.

AB. A flower having partially connected nectar.

Ab—. Latin prefix which means 'from'.

ABA. Abscisic acid.

Abaca (Manila hemp). The strongest of the hard, natural fibres. It is extracted from the pseudostems of Musa textilis, a relative of the edible bananas. Grown mainly in the Philippines, and native to that region, it is used to make ropes that are resistant to humidity and to salt water and fresh water, and to make specialist strong papers (*e.g.* for tea bags).

Abactinal. The surface of the body opposite to the mouth in an animal that is having radial symmetry.

Abalone. A single-shelled marine mollusc related to limpets and also known as an ormer or earshell. The body is sanil-like and fringed with tentacles. The shell has a line of holes across it, through which water is exhaled. Some abalones are among the largest of shellfish. They live in many parts of the world but are commonest in warmer waters.

A-bands. A-Discs. Refer to dark bands which are running transversely across the fibres of striped muscle. They are formed by filaments of the protein called myosin partly interlocking with thinner filaments of another protein called actin, the latter forming the so-called I-discs.

Abaxial

1. *zool.* Outside of or directed away from the main or central axis.

2. *bot. a.* Facing away from the stem, relative to a leaf surface. *b.* Lying outside a seed axis, relative to an embryo.

Abdomen

1. In vertebrates, the region of the body that contains the internal organs other than the heart and lungs. In Mammalia it is bounded anteriorly by the diaphragm.

2. In most arthropods, the hind region (tagma) of the body, which contains most of the digestive tract, the gonads, and the gential openings. In Crustacea, the abdomen bears limbs which are to a greater or lesser extent segmentally arranged and the abdomen is not homologous with that of arachnids and insects. The abdomen usually shows at least some trace of segmentation, though in the course of evolution this has been lost in all but one family of spiders.

3. In insects, the segments of the body that lie posterior to the thorax. The abdominal segments carry no limbs, although there are appendages (associated with

reproduction) on the terminal segments. In the primitive state the abdomen consists of eleven segments, but this number may be very much reduced in advanced insects.

Abdominal Fins. In fish, pelvic fins located far back on the belly of a fish, rather than in the thoracic or jugular position. The term may also be applied to ventral fins found on the abdominal (or belly) side of the body of a fish.

Abducens. In vertebrates, the sixth cranial nerve consisting principally of motor fibers terminating in the lateral rectus muscle of the eye.

Abducens Nerve. (Cranial nerve VI). One of the pair of nerves arising from the anterior end of the medulla oblongata in the vertebrate brain to supply the posterior rectus muscle of each eyeball. It consists mainly of motor nerve fibres. It serves to rotate the eyeball.

Abduct. To draw away from a median plane or axis.

Abductor. A muscle which is able to pull a limb away from the main axis of the body.

Aberrant. Deviating from type.

Abies (fir; family Pinaceae). Genus of coniferous trees. The leaves are crowded on the twigs, often approximately in two rows; they are needle-like and single, leaving a round, flat scar when they fall, so giving a smooth twig. The female cones are borne erect : they shatter at maturity, but the woody axis persists. Many fine ornamental species are now widely cultivated for their lofty, deeply pyramidal, monopodial crown; but they require a moist climate and are sensitive to atmospheric pollution. There are 40—50 species, widespread throughout the northern hemisphere, especially on mountains.

Abioseston. Nonliving components of the section : Tripton.

Abiotic. Non-living; devoid of life.

Abiotic Factors. The nonliving components of the environment that directly affect plant and animal life, such as water, carbon dioxide, oxygen, and light. Abiotic factors include climatic, edaphic, and physiographic factors. Compare biotic factors.

Abiogenesis. The development of living organisms from nonliving matter as envisaged in the modern theory of the oringin of life.

Abjection. The sending off of spores by the sporophore of a fungus.

Ablastin. An antibodylike substance in rats which inhibits the multiplication of trypanosomes.

Ablation

1. Removal of snow and ice by melting and direct alteration from the solid to gaseous phase (sublimation). The rate of loss is controlled chiefly by air temperature, wind velocity, humidity, rainfall, and solar radiation. Ablation on snowfields is also influenced by aspect, depth of snow, and nature of the underlying surface. Ablation till (flow till) is the glacial debris that may be released. The ablation zone of a glacier is that area in which losses, including claving, exceed additions.

2. Removal of rock material, especially by wind action.

Abomasum. The fourth and last region of the specialized stomach of ruminants (*e.g.* the cow), sometimes called the rennet stomach. It is lined with normal mucosa having gastric pits, producing hydrochloric acid, pepsin, and rennin, and is the true digestive stomach.

Aboral. Refers to the side of an animal which is away from the mouth in those animals that do not have definite dorsal and ventral surfaces, or whose dorsal and ventral surfaces change position in the course of development (*e.g.* in a starfish).

Abort

1. To miscarry or give birth prematurely.
2. To check or fall short of full development.
3. To remain rudimentary or fail to reach maturity.

Abortion. (Of part of plant or animal) condition of being stopped in normal development.

ABO System. Refers to one of the most important human blood group systems. The system has been based on the presence or absence of antigens A and B on the surface of red blood cells and antibodies against these in blood serum. A person whose blood is having either or both these antibodies cannot receive a transfusion of blood having the corresponding antigens as this would make the red cells to clumb (see agglutination). The table gives the basis of the system : people of blood group O are known as 'universal donors' because they can give blood to those of any of the other groups.

ABO Blood Group Systems

Group	*Antigens on red cell surface*	*Antibodies in serum*	*Blood group of people donor can receive blood from*	*Blood Group of people donor can give blood to*
A	A	anti-B	A,O	A,AB
B	B	anti-A	B,O	B,AB
AB	A and B	none	A, BAB, O	AB
O	neither A nor B	anti-A and anti-B	O	A,B,AB,O

Abranchiate. Without gills.

Abrasion. The erosive action that occurs when rock particles of varying size are dragged over or hurled against a surface. Some common agents of abrasion are the bedloads of streams, rock debris embedded in the bases of glaciers, and sand and shingle carried by wind or waves.

Abrocomidae. (rat chinchillas; order Rodentia, suborder Hystricomorpha). Family of mediumsized, rat-like rodents, with stiff hairs projecting over the nails of the three central digits of the hind feet. The skull is massize and narrow in the facial region; the braincase is rounded, with low occipital crests and short paroccipital processes; and the cheek teeth are hypsodont. Rat chinchillas live in colonies in burrows or crevices, and can climb trees. They are found only in the Andes from S. Peru to N. Argentina. There are two species in a single genus, Abrocoma.

Abrupt. Terminating suddenly as though cut off.

Abruptly Pinnate. (of leaves) Pinnate with no small leaf ending the stem, opposite odd pinnate.

Abscissa-phelloid. The unsuberised cells of an abscission layer in the bark.

Abscisic Acid (ABA, dormin). A growth substance that exerts numerous different mainly inhibitory, effects on the growth and development, of many species. It is a sesquiterpenoid, formula $C_{15}H_{20}O_4$, (diagram) and, like gibberellin, may be synthesized from mevalonic acid. It is active, possibly in association with gibberellic acid, in the promotion of leaf and fruit abscission and the control of dormancy. It prevents cell elongation and shoot growth and also inhibits seed germination and some tropic responses. At physiological concentrations, ABA is not toxic to plants, which can, if necessary, remove its effects by converting it to an abscisyl glucoside by linking it to a glucose residue. A large proportion of ABA is synthesized in the chloroplasts. The rate of synthesis increases dramatically who the plant is under stress, especially from water shortage. ABA overrides the normal diurnal pattern of stomatal opening and closure and causes the stomata to close during the day. This response decreases water loss by transpiration in times of drought.

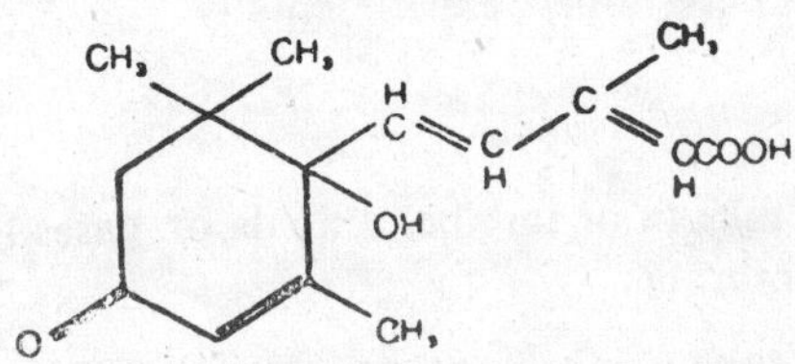

Abscisic acid

Abscission. Rejection of plant organs, *e.g.* of leaves in autumn. This occurs at an abscission zone, where hydrolytic enzymes reduce cell adhesion. The process can be inhibited by respiratory poisons, and is controlled in nature by the proportion and gradients of auxin and ethene. Other hormones may be involved.

Abscission Layer. Layer present at base of leaf-stalk in woody dicotyledons and gymnosperms, in which the parenchyma cells get separated from one another through dissolution of the middle lamella before there occurs leaf fall.

Absolute

1. Free from impurities or imperfections; perfect.
2. Not related to other factors, as absolute growth.

Absolute Pollen Frequency (APF). The expression of pollen data from sediments in terms of the absolute numbers—for each species, genus or family—per unit volume of sediment and, where deposition rates are known, per unit time. In certain circumstances this approach gives clearer information than does the traditional expression as relative pollen frequencies (RPF). APFs are particularly useful in site comparisons in which one or more high-pollen producers vary. For example, when trees first appear in the regional pollen rain their prolific pollen may, in an RPF method, give the impression of declining herbaceous species, whereas examination by an APF method will show constant values for herb species.

Absolute Transpiration. The rate of water loss from a plant which is determined experimentally.

Absorb

1. To take in or imbibe, a fluids or gases through a cell membrane.
2. To take a substance, as water or nutrients, into the body through the skin or mucous membranes, or, in plants, through root hairs.
3. To take in and stop radiant energy, as heat and ultraviolet rays.

Absorption

1. The uptake into a plant of water, solutes, or other substances by either active or passive means. Entry invariably involves movement across cellular membranes. Active absorption, for example the uptake of a solute against an osmotic gradient, involves expenditure of energy. An example of passive absorption is the intake of water by plant roots, which is controlled by the rate of transpiration.

2. The retention of radient energy by the pigments of a plant. About 80% of the incident visible light and about 10% of the infrared radiation falling on a leaf is absorbed. Generally, less than 2% of this is used in photosynthesis.

Absorption Spectrum. Used for a plot of the absorbance by a substance of radiation at different wavelengths, usually of ultraviolet, visible, or infrared radiation. It provides useful information about the identity or quantity of the substance.

Absorption Ratio. It is the ratio of the final internal concentration of a particular ion in plant tissue (*i*) to the final external concentration of a particular (*e*), *i.e.*, *i*/*e*.

Abstriction. The separation and discharge of a part as in the formation of spores or conidia in various ascomycetes.

Abutilon (family Malvaceae, tribe Abutileae). Genus of trees and shrubs, native to all arid zones of the world. It includes Chinese lanterns, grown as ornamentals. They have hairy leaves and branches, and bright, lantern-like flowers, mainly orange or red. The fruit is a schizocarp and the stigmas are usually apical. There are 30 species, all in N. Australia. The genus includes A. avicenae (China jute).

Abutilon Avicenae (China jute). Abutilon.

Abyssal Fish. Fish living in the abyssal zone of the ocean. Many abyssal species have a prominent snout; a tapering, rat-tailed body, consisting of flabby, watery tissue; and a lightweight skeleton.

Abyssal Hills. Relatively small topographical features of a dominantly flat deep-ocean floor, commonly 50—250m in height and a few kilometres in width. They are most typical of the Pacific Ocean floor at depths of 3000—6000 m.

Abyssal Plain. Smooth, almost level area of the deep-ocean floor in which the gradient is likely to be as low as 1 : 10 000. The covering sediments are usually thin deposits of a pelagic ooze or distal turbidite.

Abyssal Realm. The portion of the marine realm below the level to which light penetrates, including all waters below a level of 600 feet.

Abyssal Zone. Zone of greatest ocean depth—that is, below a depth of 2000 m. This zone lies seaward of, and deeper than, the bathyal zone, and covers approximately 75% of the total ocean floor. It is the most extensive Earth environment, cold, dark, and still, supporting fauna that typically are black or grey, delicately structured, and not streamlined.

Abyssal organisms are adapted for living under high pressures in cold dark conditions.

Abyssinian Cat. Cat.

Abyssobenthos. All organisms of the abyssal zone.

Acacia. A large group of Evergreen trees and shrubs, found mainly in the warm and dry parts of the world, including the African savanna and the Australian bush country, where they are called wattles. Many are thorny and bear cylindrical or globular clusters of yellow or white flowers. Some are known as mimosas.

Acacia, False. Locust Tree.

Acanaceous. Having sharp points.

Acantha. A sharp point.

Acanthaceous. Possessing or bearing spines.

Acanthella. A stage following the acanthor.

Acanthocephalan. A parasitic, spiny-headed worm of the phylum Acanthocephala inhabiting the intestines of vertebrates. Immature stages develop in arthropods.

Acanthoid. Spinelike.

Acanthor. The first larval stage of an acanthocephalan usually found in insect larvae.

Acanthus. Any of a group of plants with spikes of white or purple flawers and large toothed or spiky leaves. They are found in Africa, Asia and Europe, and like dry climates.

Aceana (family Rosaceae, subfamily Rosoideae, tribe Sanguisorbeae). Genus of wind-pollinated, mat-forming plants occurring mainly in the sub-tropics and temperate regions.

There are rather more than 100 species; 90 species are found in S. America, 14 in New Zealand, two in Australia, and others in New Guinea, S. Africa and California.

Acalypha (family Euphorbiaceae). Genus of shrubs and small trees. The flowers are unisexual and apetalous, males having elongate, twisted antherlobes, and females much-branched stigmas; some species bear catkins, as in the ornamental A. hispida ('cat's tail'). A. wilkesiana and others have colourful foliage. There are no close relatives in the family; the resemblance to Urticaceae is only superficial. There are about 450 species, found in the tropics and subtropics.

Acanthaceae. Family mainly of shrubs, in which the leaves are simple, opposite, and decussate. The flowers are bisexual and zygomorphic, and the bracts are often showy. There are four or five fused sepals and petals; 2—4 epipetalous stamens; a superior, bilocular ovary; and numerous, axile ovules. The fruit is a capsule. There are several ornamentals, *e.g.* Acanthus, Aphelandra, Thunbergia. There are about 250 genera, comprising about 2000 species, most of them tropical, but with temperate outliers.

Acanthiza (thornbills; family Acanthizidae). Genus of warblers, sometimes placed in the Sylviidae or Maluridae. They are small, active, insectivorous birds, usually found in small parties feeding in or close to trees and bushes. They are mainly brown with differing rump colours, and are best distinguished by their calls and behaviour. The sexes are similar. There are 11 species, found only in Australia.

Acanthizidae (bristlebirds, scrub wrens, fairy warblers, thorbills, whitefaces; order Passeriformers). Family of birds with thin, pointed bills with basal bristles, all of which build domed nests. They are insectivorous and vary from strictly arboreal (fairy warblers) to ground-feeding (scrubwrens). The sexes are usually similar. There are 17 genera, comprising 63 species. They are found mainly in Australia and New Guinea.

Acanthobdellida (phylum Annelida, class Hirudinea). Order of parasitic annelid worms that lack an anterior sucker. Bristles occur in the antreior region. There is one genus, Acanthobdella, that occurs only on salmon.

Acanthocephala. Phylum of bilaterally symmetrical, worm-like organisms with elongate bodies which may show poor segmentation. Most do not have an excretory system, and all lack a gut. A retractable proboscis is present. They are all endoparasitic, living in two or three marine or terrestrial vertebrates during their development. They phylum contains about 600 species.

Acanthochitonina (class Amphineura, subclass Polyplacophora, order Neoloricata). A suborder which contains the largest of all polyplacophorans, but small members also occur. Posterior valves may not be present. The girdle may envelop the valves totally. The first appeared in the Oligocene.

Acanthoclinidae (class Osteichthyes, subclass Actinopterygii, order Perciformes). Family of small (8 cm), agile fish inhabiting tropical waters of the Indo-Pacific region. They have pelvic fins reduced to one spine and two soft rays.

Acanthodii. Class of primitive, fossil fish, characterized by the presence of a true bony skeleton, a heterocercal tail fin, a persistent notochord, ganoid scales, and stout spines in front of the fins. The Acanthodians lived from the Silurain to the Permian Period and may be related to ancestors of the more modern bony fish.

Acanthopterygii. Superorder of bony fish including all spiny-finned fish (*e.g.* Perciformes and Beryciformes).

Acanthuridae (surgeonfish, doctorfish; class Osteichthyes, subclass Actinopterygii, order Perciformers) Large family or slab-sided fish with a sharp spine on each side of the tailbase (hence the name 'surgeonfish'). The spine is retractable in some species. Many species exhibit striking colour patterns. There are about 80 species, found in all tropical seas, often feeding on organisms covering the surface of rocks, or on seagrasses.

Acanthus (family Acanthaceae). Genus of shrubs, most of which are xeromorphic and have spiny leaves. The upper lip of one corolla is lacking. The pattern for the decoration on Corinthian column capitals is supposedly based on the leaves of A. spinosus. There are 50 species, found in S. Europe (A. mollis is bear's breeches), Asia, and Africa (A. ilicifolius grows mangrove swamps).

Acarid. Any arachnid of the order Acarina, which includes the ticks and mites.

Acarina (acari mites, ticks; class Arachnida). Order of small or very small arachnids that have a short, unsegmented abdomen. The body comprises three tagmata : the gnathosoma, the propodosoma (these two sometimes jointly called the protosoma), and the hysterosoma. The respiratory organs are tracheae. Many species are parasites of medical and veterinary importance, and many are agricultural pests. The Acarina are regarded by many zoologists as an unnatural, polyphyletic group comprising elements derived independently from different arachnid stocks. There are 20 000 known species, 2800 of which are aquatic, including some that are marine.

Acarinum. A small pouch in the abdomen of Old World carpenter bees of the genus Xylocopa (subgenus Koptortosoma), which provide protection for symbiotic mites of the genus Dinogamasus.

Acarpellous. Having no carpels.

Acarpotropic. Not shedding its fruit.

Acarp ous. Producing no fruit.

Acaulescent. To describe plants which do not have stem or which have an extremely short stem such as tufted or rosette plants.

Acaulous, -ine, One. Having no true stem or only a very short one under the earth.

Accentor. A member of a small family of birds found in Europe and Asia. The commonest species is the dunnock, also called the hedge sparrow, although it is not a sparrow at all. Accentors have slender beaks and feed on insects and small seeds.

Acceptor. A substance which accepts or combines with a product of a chemical reaction, especially a substance such as cytochrome or a vitamin which accepts and transfers hydrogen atoms in cellular respiration. Compare donor.

Accessory. Accompanying as a subordinate; aiding or contributing in a secondary way; additional.

Accessory Bud. A bud which is present additional to the normal axillary bud.

Accessory Cell. A cell which is associated with the guard cell of a stoma, but is differing in structure from it and from the ordinary cells of the epidermis.

Accessory Character. A characteristic of a species which is not essential to its diagonosis, but is sometimes employed to differentiate races.

Accessory Chromosome. Sex Chromosome.

Accessory Fructification. An asexual reproductive structure of lower plants.

Accessory Genitalia (secondary genitalia). Organ of intromission, present only in the males of the order Odonata. It is situated on the sternites of the second and third abdominal segments. It has no homologues in the animal kingdom.

Accessory Multiplication. Any asexual reproductive process.

Accessory Nerve (Cranial nerve XI). One of the pair of nerves arising from the posterior region of the medulla oblongata in the brain of higherver tebrates. It carries motor have fibres and merges with the adjacent vagus nerve close to its root. In fishes and amphibians it is regarded as an integral part of the vagus.

Accessory Pigments Pigments other than chlorophyll *a* found in photosynthetic cells. They include the carotenes and the xanthophylls, together known as the carotenoids, chlorophylls *b*, *c*, and *d*, and the phycobilins. The latter are found only in blue-green and red algae. The composition of accessory pigments in algae is used as a taxonomic character.

The accessory pigments function as secondary absorbers of light in regions of the visible spectrum not covered by chlorophyll *a*. The light energy that they absorb must be transferred to chlorophyll *a* before it can be used in the photosynthetic process. As energy transfer from one molecule to another can only occur from a shorter wavelength absorbing form to a longer wavelength absorbing form, all accessory pigments have absorption maxima at shorter wavelengths than chlorophyll *a*. Energy is passed from chlorophyll *a* to the reaction centre pigments, P680 and P700.

Some accessory pigments may have a protective function, preventing photooxidation of the cell's chlorophyll at high light intensities.

Accessory Respiratory Organ. A system of air chambers formed by outgrowths from the mouth or gill region of fish that occasionally leave their aquatic environment. The uptake of oxygen from the air is facilitated by a dense network of tiny blood vessels in the skin that lines these air chambers. Possession of these organs enables such fish as labyrinthfish (Anabantidae), snakeheads (Channidae), or air-breathing catfish (Clariidae) to survive outside water for some considerable time. The swim-bladder also may serve as an accessory respiratory organ.

Accessory Sex Organs. Organs assisting in the process of sexual reproduction, *e.g.*, the nuptial pads of male frogs, claspers of insects etc.

Accessory Spore. An asexual spore.

Accessory Transfusion Tissue. Transfusion tissue which is extending through the mesophyll from the mid rib to the margins, found in certain gymnosperms, *e.g.*, Cycas.

Accidental Species. One of five classes of fidelity used by the Braun-blanquet school of phytosociology in the description and classfication of plant communities. Accidentals are rare species in the community, present either as chance invades from another community or as relics from a previous community. Compare exclusive species; indifferent species; preferential species; and selective species.

Accipiter (sparrowhawks, goshawks; family Accipitride). The largest genus of birds of prey, comprising small to medium-sized hawks. Broad wings and long tails enable them to pursue prey through the woodland in which they live, and where they feed on mammals, birds, reptiles, and insects. The females are often larger than the males. There are 47 species, whose distribution is cosmopolitan.

Accipitridae (hawks, eagles, buzzards, kites, Old World vultures; order Falconiformes). The largest family of birds of prey, with many genera cosmopolitan in their distribution, especially Accipiter, Buteo, Circus, Elanus, and Haliaeetus, which make up nearly half the family. Other genera are more restricted in range, and more than half are monotypic. They are generally carnivorous, preying on snails, insects, fish, mammals, reptiles, and birds; but one species, Gypohierax angolensis (palm-nut vulture), feeds on oil-palm husks. There are 63 genera, with more than 200 species.

Acclimation. Response by an animal that enables it to tolerate a change in a single factor (*e.g.*, temperature) in its environment. Applied most commonly to animals used in laboratory experiments.

Acclimatisation Used for describing adjustment of an animal organism with some particular condition of its environment by making necessary physical, physiological, biochemical and behavioural changes *e.g.*, as frog being cold blooded animal it can not withstand excessive hot or cold weather and therefore during winters it undergoes hibernation which involves lowering of BMR, consequent suspended animation thereby minimising the vital needs of the body to cope with the conditions of cold while during summers it undergoes aestivation.

Accommodation

1. (in animal physiology). The process by which the focal length of the lens of the eye is changed so that clear images of objects at a range of distances are displayed on the retina. In man and some other mammals accommodation is achieved by reflex adjustments in the shape of the lens brought about by relaxation and contraction of muscles within the ciliary body.
2. (in botany). The ability of a plant to adapt itself to gradually changing environmental conditions.
3. (in animal behaviour). The psychological adjustment made by an animal in response to continuously changing environmental conditions.

Accrescent. Enlarged and persistent. Especially of a calyx which gets enlarged as the fruit ripens.

Accretion. Growth or increase in size by external addition.

Accumatization. The act or process of becoming adjusted to or tolerant of a new and different climate.

Accumbent. Touching something specially of cotyledons having edges against radicle.

Accumulated Temperature. Surplus or deficit of temperature with respect to a defined mean value and expressed as an accumulation over a given period, *e.g*, month, season, or year. For example, a datum value of 6°C is used as a critical temperature for sustained vegetation growth.

Accumulator. Term used in plant-succession studies to describe a pioneer species whose activities are claimed to enrich the abiotic environment with nutrients.

Aceae. Standardized suffix used to indicate a ranking of plants at the family level in the recognized code of classification, *e.g.*, in 'Rosaceae', the rose family.

Acellular. Describing tissues, organs, or organisms consisting of a mass of protoplasm not divided by cell walls into discrete units. The term may be used of relatively large uninucleate organisms, such as the alga Acetabularia, to distinguish them from comparatively less specialized and smaller unicellular organisms. More often acellular structures are multinucleate and result from free nuclear division with no accompanying cell wall formation. When describing multinucleate organisms, tissues, or cells the term is synonymous with coenocytic. Examples of acellular multinucleate tissues are the endosperm of certain angiosperms and the initial stages in formation of the proembryo in Cycas. Acellular multinucleate organisms include algae of the order Siphonales.

Acellular Slime Moulds. Slime moulds in which the feeding stage is a true plasmodium.

Acentric. Applied to a fragment of a chromosome, formed during cell division, that lacks a centromere. The fragment will be unable to follow the rest of the chromosome in migration towards one or other pole as it has lost its point of attachment to the spindle.

Acephalous. Without a head. The term is also referred for such creatures whose heads are much reduced or are invaginated into the thorax, as in certain insect larvae.

Acer (family Aceraceae). Genus of trees in which the leaves are without stipules, opposite, and entire or often palmately lobed, but rarely pinnate (*e.g.*,) they are in A. negundo, box elder). Regular, pentamerous flowers are borne in racemes, corymbs, or umbels. The fruit is a samara. These trees are important for timber; many are ornamentals; and maple syrup is obtained mainly from A. saccharum. Field maple (A. campestre) is wild in Britain; sycamore (A pseudoplatanus), from S. Europe, is naturalized in Britain. There are about 200 species, mainly northern temperate, and a few in tropical mountains. China has 100—150 species.

Aceraceae. Small family of two genera of trees, Acer and Dipteronia (which comprises two species found only in China). The family is related to Hippocastanaceae and Sapindaceae.

Acerose. Sharp-pointed; needlelike.

Acerous. Lacking horns, tentacles, or antennae.

Acervate. Heaped; growth in heaps or groups.

Acervulus. A small disc-shaped mass of conidiophores that erupts through the epidermis of plants infected by fungi of the order Melanconiales.

Acetabularia (mermaid's cup, mermaid's wineglass; family Dasycladaceae, order Dasycladales). Genus of algae in which the mature thallus consists of a single axis or 'stalk' with a gametangial disc at the top. Found in warm regions in shallow, sheltered seas, attached to rocks, shells, etc. Fossil Acetabularia dating from the Tertiary have been found.

Acetabuliform. Saucer-like.

Acetabulum

1. The cup-shaped socket on the lateral surface of the hipbone of tetrapods.
2. The ventral sucker of a trematode.
3. The genital sucker of a time.

Acetification. The production of vinegar (acetic acid, CH_3COOH) from alcohol by aerobic, acetic acid bacteria of the genus Acetobacter.

Acetaldehyde (ethanal). An intermediate in the conversion of pyruvic acid to ethanol, the final stage of glycolysis under anaerobic conditions in plants. The conversion of acetaldehyde to carbon dioxide and enthanol is energy requiring, involving oxidation of a reduced molecule of NAD. Acetaldehyde is also involved in the synthesis and breakdown of the amino acid threonine.

Acetic Acid (ethanoic acid). A weak organic acid, formula CH_3COOH. Acetic acid can be used as an alternative carbon source by certain algae (*e.g.* Chlamydomonas mundana and species of (Chlorella), white green algas the genus Chlamydobotrys are totally dependent on acetic acid as a carbon source. In combination with coenzyme A (acetyl CoA) acetic acid plays a central role in aerobic energy metabolism.

Acetic-acid Bacterium. Bacterium that produces acetic acid from ethanol (ethyl alcohol).

Acetobacter. Genus of Gram-negative, aerobic bacteria not included in any taxonomic family. Cells are ovoid or rod-shaped, motile or non-motile. Certain species can oxidized ethanol (ethyl alcohol) to carbon dioxide, forming acetic acid as an intermediate. Some species are used in the manufacture of vinegar. They are found on fruits and vegetables, in alcoholic beverages, etc.

Acetone. Ketone body.

Aceto-orcein. A stain used in the preparation of root-tip or another squashes for chromosome examination. The material to be stained is placed in the aceto-orcein (which is acidified with hydrochloric acid) and heated at 60°C for 15 minutes. The material is then removed and mounted in acetic acid.

Acetylation. The formation of an acetyl radical from pyruvate during photosynthesis. The enzyme involved is co-enzyme

A. The complete reaction may be put as follows :

(1) $CH_3.CO.COO^- + CoA + 2H_2O \longrightarrow$
Pyruvate
$CH_3.COO.CoA + HOC_3^- + H^+$
Acetyl CoA

(2) Acetyl CoA + oxalacetate ⟶ citrate + CoA

Acetylcholine. A chemical substance ($CH_3CO.O.CH_2CH_2N(CH_3)_3 OH$) which is produced at the synapses and nerve endings of the parasympathetic nerves. Its effects are the exact opposite to those which are produced by sympathin, a substance similar to adrenalin, which is produced by the sympathetic nerves. Acetylcholine for example, slows down the heart while sympathin speed it up.

Ac ylcholinesterase. Refers to an enzyme hydrolysing acetlycholine into choline and acetate; present on the surface of post-synaptic membrane. Enzyme molecule is having two distinct sites : an anionic site which binds the quaternary nitrogen of acetylcholine and an esteratic site which can donote electron to the acetate portion of acetylacholine molecule separating it from the choline moiety. This permits the hydrolysis of acetylcholine and terminates the post-synaptic action of the transmitter.

Acetyl CoA (acetyl coenzyme A). A compound consisting of acetyl combined through a sulphur bridge with coenzyme A. The formation of acetly CoA is an energy-requiring reaction, involving the conversion of ATP to AMP and pyrophosphate. Acetyl COA plays a central role in intermediary metabolism. It is a product of the degradation of fatty acids, carbohydrates, and some amino acids. It is an essential precursor in the TCA and glyoxylate cycles and is the starting point for the synthesis of fatty acids, terpenes, and some amino acids.

Acetyl Coenzyme A (Acetyl Co-A). A compound of key importance in intermediary metabolism because it is having higher energy content than acetic acid. It reacts readily with substances in reactions which would otherwise need outside energy. From energetics point of view, formation of 1 mol of compound is equal to the formation of 1 mol of ATP. Formed by pyruvic acid in Kreb's cycle and also by β-oxidation of fatty acids.

N-Acetyl-2-Glucosamine. A glucose derivative, and the unit which gets repeated in the chitin molecule.

Acetyl-S-CoA. An intermediate, which is produced by co-enzyme A when it gets detached carbon dioxide from the α-keto-acid during respiration.

Achaetous or Achetous. Without setae or bristles.

Achane or Akene. A simple, one-seeded, indehiscent fruit in which the seed is attached to the pericarp at only one point, as in the sunflower, buttercup, and sycamore.

Achene. A dry indehiscent fruit formed from a single carpel and containing a single seed. An example is the feathery achene of clematis. Variants of the achene include the caryopsis, cypsela, nut and samara.

Acherontia Atropos. (death' head hawkmoth; order Lepidoptera, family Sphingidae). Very large moth whose common name is derived from the skull-like pattern on the thorax. It robs beehives of honey. The moth squeaks by forcibly expelling air through its proboscis. The larva is very large, green or purple, and feeds on Solanaceae. The species is found in Africa, Europe, and N. Asia.

Achilary. Lacking a lip.

Achiamydeous. Lacking a calyx or corolla.

Acholeplasmataceae (class Mollicutes). Family of Gram-negative, prokaryotic, chemo-organotrophic organisms that do not require the presence of sterols for growth. Cells are spherical, pleomorphic, or filamentous. They are found as parasites in a variety of mammals and birds. There is one genus (Acholeplasma), containing five species.

Achras (family Sapotaceae). Genus of plants. A. zapota (Manilkara zapota) yields the delicious fruit chiku, or sapodilla plum, for which it is now widely cultivated; and chicle, formerly the elastic component of chewing gum. There are four species, found in northern tropical America.

Achromatic

1. Lacking in color.
2. Not staining readily, with reference to tissues.
3. Refracting light without decomposing it into its constituent colors, with reference to lenses.

Achromatic Apparatis. In mitosis the spindle fibers and cell centers, which do not stain readily. Also called achromatic figure.

Achromatic Figure. The metaphase plate chromosomal line up.

Achromatic Spindle. ('Spindle'). In cell division, a system of apparent fibres which does not get stained by basic dyes. They join the poles of the nucleus, and get diverged towards the centre of the cell.

Achilles, Tendon. The tendon of the heel : a strong tendon connecting the gastrocnemius and solaeus muscles to the calcaneum in terapods.

Acicula pl. Aciculae. A needlelike spine or bristle.

Acicular. Needle shaped, for example the acicular crystals that, closely packed together, form a raphide, and the leavs of pine (Pinus) trees.

Aciculum. Refers to an internal needle-like chaeta or bristle which gives support to the external chaetae on certain polychaete worms. Free-swimming polychaetes such as

Nereis are having on each segment a pair of flap-like projections called parapodia on which there are a number of external chaete.

Acid. According to the Bronsted-lowry theory, a substance that in solution liberates hydrogen ions or protons. The Lewis theory states that it is a substance that acts as an electron-pair acceptor. An acid reacts with a base to give a salt and water (neutralization), and has a pH of less than 7.

Acid-base Balance. The regulation of the concentrations of acids and bases in blood and other body fluids so that the pH remains within a physiologically acceptabe range. This is achieved by the presence of natural buffer systems, such as the haemoglobin, bicarbonate ions, and carbonic acid in mammalian blood. By acting in conjunction, these effectively mop up excess acids and bases and therefore prevent any large shifts in blood pH. The acid-base balance is also influenced by the selective removal of certain ions by the kidneys and the rate of removal of carbon dioxide from the lungs.

Acid Dyes. Dyes having an acidic organic grouping of atoms (anion) which is the actively staining patr, combined with an inorganic cation, *e.g.*, eosin. These dyes stain cytoplasm and collagen.

Acid-fast Bacteria. Certain bacteria of the order Action my cetea a including Mycobacterium which after being treated with certain dyes, are not decolorized on subsequent treatmant with a mineral acid.

Acid-fast Stain. A widely applied stain in bacteriology that is used to identify organisms that can retain a dye on washing with acid alcohol. One technique is to stain a bacterial smear with hot carbol fuchsin and, after rinsing with water, expose the smear to concentrated hydrochloric acid dissolved in water or ethanol. Following a second rinsing, the smear is counterstained with methylene blue. Acid-fast organisms retain the red colour of the carbol fuchsin while other organisms appear blue. Examples of acid-fast bacteria are species of the genus Mycobacterium.

Acidic Grassland. Grassland that occurs on acidic soils : it is usually derived from former woodland as a consequence of centuries of grazing and, to a lesser extent, burning. In Britain and much of NW Europe the dominant grasses are species of Agrostis (bent) and Festuca (fescue). This type of vegetation is most extensive in upland areas, but the associated plant species tend to be different, and the name 'grass heath' is considered more appropriate.

Acidic Stains. Staining.

Acidophilic

1. Refers to the propensity of a cell, its components, or its products to become stained by an acidic dye.
2. Acidophilous.

Acidophilic. Staining readily with acid stains, as eosin.

Acidophilous (acidophilic). Applied to 'acid-loving' organisms, *i e.* organisms that grow best in acidic habitats.

Acidopore. Formicinae.

Acid Rain. Rain with a high level of acidity caused by pollution by oxides of nitrogen and sulphur produced by coal and oil combustion. The oxides combine with water in the atmosphere forming acids such as nitric and sulphuric acids. Acid rain has caused the deaths of countless living organisms both by producing a high level of acidity and by leaching metal ions from the soil into rivers and lakes. The disastrous effects have been extensively documented in Scandinavia and North America. Recently, the Warren Spring preliminary report has revealed a similar level of pollution in Scotland and parts of northern England.

Acid Rock. Igneous rock containing more than about 60% silica by weight, most of the silica being in the form of silicate minerals, but with the excess of about 10% as free quartz. Typical acid rocks are granites, granodiorites, and rhyolites'. The limits are somewhat arbitrary and there are a number a schemes.

Acid Soil. Soil having a pH less than 7.0. Degrees of soil acidity are recognized. Soil is regarded as 'very acid' when the reaction is less than pH 5.0. The USDA lists five standard ranges of soil acidity (less than pH 4.5, extremely acid; 4.5—0.5, very strongly acid; 5.1—5.5, strongly acid; 5.6—6.0, medium acid; and 6.1—6.5, slightly acid). Surface soil horizons of acid brown earths have a reaction of pH 0.5 or less.

Acid Stain. Staining.

Acinar. Composed of acini, as an acinar gland; comprising acini, as acinar cells.

Acinonyx Jubatus (cheetah). Felidae.

Acinus pl **Acini.** A saclike secreting unit of a gland, especially one with a narrow lumen.

Acipenseridae (sturgeons; class Osteichthyes, subclass Actinopterygii, order Acipenseriformes). Family of large, fairly sluggish, bottom-feeding fish with a low-slung mouth, toothless jaws, four barbels in front of the upper jaw, and rows of large plates along the body. Found in Europe and N. America. The majority of the 26 species are marine but ascend rivers to spawn. A few species, *e.g.* the sterlet (Acipenser ruthenus), live in fresh water, never reaching the sea. Sturgeons are very fertile, large females of the common sturgeon (Acipenser sturio) producing up to 6 000 000 eggs per year. The roe of ripe females after processing is renowned as caviar.

Acipenseriformes (class Osteichthyes, subclass Actinopterygii, superorder Chondrostei). Order of rather primitive fish characterized by a cartilaginous skeleton, a heterocercal tail-fin, and the head showing a pointy, protruding rostrum with a ventrally located mouth. Extant members of the order either have a naked skin or possess five rows of large, rhomboid, bony scales along the body; a spiral valve is found in the intestinal tract. Apart from the sturgeon and the paddlefish the order includes a number of fossil representatives dating back to the Carboniferous Period.

Acme. The period of maximum vigour of an individual, race, or species.

Acoela

1. (class Turbellaria, subclass Archoophora). Order of platyhelminth worms that have no gut cavity or pharynx.

2. (phylum Mollusca, class Gastropoda). Order of opisthobranch gastropods that possess no shell, mantle cavity, or gills. Respiration is carried out by branchiae. Dorsal outgrowths are quite common, and all have undergone complete detorsion. Benthic and planktonic forms occur. There is only one fossil family recorded, which appeared in the Eocene. According to some authorities, the taxonomic use of 'Acoela' as a molluscan order is little accepted and should be discouraged. It represents an attempt to unite the Notaspidea and Nudibranchia.

Acoelomate. Lacking a coelom.

Acoelomate. Any member of the Acoelomata, a group of metazoans comprising the phyla Porifera, Coelenterata, and Ctenophora characterized by lack of a coelum.

Acoeloraphe (formerly Pourotis; family Palmae). Monotypic palm genus. A. wrightii is a low, clump-forming, monoecious, fan palm with greyish leaves, native to the Florida Everglades, and widely cultivated.

Acoelous. The first two centrum of a vertebra in which both the sides are convex *e.g.* IX vertebra of frog.

Aconite. Eranthis.

Aconite Winter. A low flowering plant with yellow blooms, belonging to the buttercup family. It is found mainly in gardnes.

Aconitrase. An enzyme which takes part in the Citric Acid Cycle (Kreb's Cycle). It dehydrates citrate forming *cis*-aconitrate, which is then converted to D-isocitrate, by the addition of a water molecule.

Acontium. In Anthozoa, an extension of the mesentery-bearing nematocysts which lies in the gastric cavity but which can be protruded through the mouth in order to capture prey.

Acorn. The nut or fruit of the oak tree.

Acorn Barnacle. Balenus balanoides; and Balanidae.

Acorn-shell. Barnacle.

Acorn Worm. A strange animal that lives in the mud of the seabed, often at great depths The 'acorn' is the head end of the animal, which sticks out of the mud and is bright orange or red. There are several species, ranging from about 5 cm to 2 metres in length. The body is rather like that of an earthworm at first glance, but the animal breathes by means of gills. It is something of a link between the invertebrates and the vertebrates.

Acorus (family Araceae). Genus of two species of warm temperate Asia. A. calamus (sweet flag), a marsh plant, is widely cultivated for its fragrant leaves and rhizomes with medicinal properties. The rhizomes are sympodial, the flowers bisexual.

Acotyledon. A plant without cotyledons (said of the embryo of a higher plant).

Acoustic. Auditory; pertaining to the sense of hearing or sound.

Acoustic Nerve. Vestibulocochlear nerve.

Acoustico-lateralis System. Refers to a system in vertebrates (especially fish and aquatic amphibians) which is having neuromast cells in various tracts on or near the surface of the body (*i.e.* the lateral-line system) and the inner ear (membranous labyrinth). Neuromasts and receptor cells of the inner ear are similar. They consist of groups of cells having sensory processes which are innervated by nerves from the medulla oblongata of the brain. They are nearly helpful in detection of sounds in water.

ACP. Acyl carrier protein.

Acquired

1. Developed as a result of the effects of the environment; not inherited.
2. Gained from experience and training.

Acquired Character. A trait or character which is not inherited; one which has developed as a result of the effects of the environment or from use or disuse.

Acquired Characteristic. A characteristic of a living structure or biochemical system that has been brougnt about by environmental factors acting during the life of the organism. For example, plants receiving fertilizer applications may grow taller and more vigorously than before and pot plants moved to a shady position may become etiolated. It is a fundamental tenet of Neo-darwinism that such variations are not inherited by the succeeding generation.

Acquired Variation. A structural anomaly which becomes evident as the individual gets developed.

Acrandrous. Possessing antheridia at the top of the stem, as in some mosses.

Acrania. (Protochordata). A subdivision of the Chordata including two subphyla called the Urochordata and the Cephalochordata, (*i.e.* all the invertebrate chordates). Brain, skull, heart, and kidneys are not present and the notochord is never replaced by cartilage or bone. Sometimes the term is used for the Cephalochordata only.

Acraniate. Any member of the Acraniata, a group of chordates comprising the subphyla Cephalochordata and Urochordata characterized by absence of a brain and cranium; the protochordates.

Acrasiales (subclass Acrasidae). Order of slime moulds with the characteristics of the subclass. The order contains two families and several species, but classification remains tentative. The organisms are found on dung, soil, decaying plant material, etc.

Acrasidae (class Acrasiomycetes). Subclass of slime moulds in which the myxamoebae have blunt, rounded pseudopodia. When aggregating to form a pseudoplasmodium, the myxamoebae do not collect into distinct streams. There is one order.

Acrasin. Chemotactic substance involved in the aggregation response of cellular slime moulds (Acrasiales). It is now known to be cyclic AMP.

Acrasiomycete. Any member of the class Acrasiomycetes, comprising the cellular slime molds which possess a unicellular, amoeboid assimilative stage.

Acrasiomycetes (cellular slime moulds). A class of the Myxomycota containing organisms that feed on soil bacteria. It includes some 25 species divided among 10 genera. Cellular slime moulds are made up of an aggregation of myxamoebae forming a pseudoplasmodium. These produce uncovered masses of walled spores or pseudospores by a process different from that of other fungi. The cellular slime moulds are often classified as an order (Acrasiales) of the Myxomycetes.

Acraspedote. Lacking a valum, with reference to medusae.

Acre. Unit of area, the British statute acre being equal to 0.4047 hectares or 4840 square yards. In Britain, historically a woodland acre varied from forest to forest, but was always larger than the statute measure. Assarts were surveyed at the statute measure.

Acrididae (short-horned grasshoppers, locusts; order Orthoptera, suborder (Caelifera). Cosmopolitan family of small to large, short-horned grasshoppers among which stridulation is common, the males and some females rubbing a row of pegs on the hind femora against a toughened vein of the forewing '(tegmen). They have three-segmented tarsi and antennae that are shorter than the forefemora, and hind legs modified for jumping (saltatorial). There are tympanal organs (ears) at the base of the abdomen. The female has a short ovipositor and lays eggs in the soil, or sometimes in decayed wood. All species are plant feeders and a number are important agricultural pests, most notably the locusts. Lubber, spur-throated, slantfaced, and band-winged grasshoppers are members of the family, which is the largest of the Caelifera and contains some 10 000 known species.

Acridotheres (mynahs; family Sturnidae). Genus of generally black or dark-brown birds with white at the base of the primaries and at the tip of the tail. They nest in holes in trees, or in banks. They are relatively omnivorous and all species are commonly found in captivity; Acridotheres tristis (common mynah) has become semi-domesticated and has been introduced to Australia, New Zealand, and S. Africa. There are five species, found from S. Russia and China, and south across SE Asia to Celebes.

Acrocarpic. Having the fruiting body at the tip of the main axis, as certain mosses.

Acrocarpous Moss. Type of moss in which the archegonia, and hence the capsules, are borne at the tips of stems or branches. Acrocarpous mosses usually show little or no branching and typically grow in erect tufts. Compare pleurocarpous moss.

Acrocentric. Describing a chromosome in which the centromere is not centrally placed, giving two arms of different lengths. Compare metacentric, telocentric.

Acrocephalus (reed warblers; family Sylviidae). Genus of small, brown warblers, mainly inhabiting marshes, and sometimes included in Muscicapidae. They are mainly insectivorous, but sometimes eat seeds and berries. Their nests are suspended between stems of reeds and bushes. They have distinctive, loud, chattering songs. There are 26—30 species. Many are migratory, breeding in Eurasia and wintering in Africa and SE Asia.

Acrochordidae (wart snakes; order Squamata, suborder Serpentes (Ophidia). Family of primitive snakes with ridged scales that do not overlap. There are nasal openings on top of the snout. Usually they are found in brackish water, but occasionally offshore. They rarely come on to land. There are two species, found in India, Sri Lanka, and some Indo-Australian islands.

Acrodont. Designating a type of tooth attachment in which sockets are lacking, the teeth being fused to the bone, as in Sphenodon.

Acrodromous. The veins of leaves run parallel to the edge of the leaf and meet at the tip.

Acroechinoidea (class Echinoidea). Name of a taxonomic rank between the subclass and superorder levels that comprises the monophyletic group composed of all Echinoidea except for the Echinothurioida.

Acrogamae. A division of the Angiospermae in which fertilization occurs through the micropyle (the usual way.

Acrogen. Any plant of the highest series of cryptogams, the archegoniatae.

Acrogynous. Developing at the tip of the main shoot, with reference to the archegonia of certain bryophytes. Compare anacrogynous.

Acromegaly. A disease which is caused by hypersecretion of anterior pituitary after bones have altarned full growth; bones thickened especially in the limbs and face; characterised by an overgrowth of mandibles, phalangeal certilages and related areas, forehead, nose, hands, feet etc. Body is becoming disproportionate giant.

Acromion. A combination of the spine of the scapula which is forming a projection for articulation with the clavicle in many mammals and birds.

Acron. The anterior, unsegmented head in embryonic or ancestral arthropods.

Acronematic Flagellum. Whiplash flagellum.

Acropetal. Growing or developing upwards from the base or point of attachment, so that the oldest parts are at the base and the youngest are at the tip.

Acropetal Movement (basifugal movement). Movement of substances within the ptant toward its root and shoot apices. Compare basipetal movement.

Acroscopic. Facing in the direction of apex.

Acrophilous. Inhabiting regions of high altitude.

Acropomatidae (class Osteichthyes, subclass Actinopterygii, order Perciformes) Very small family of marine tropical (Indo Pacific) fish, comprising only three species. A characteristic of this family is the anterior location, close to the pelvic-fin base, of the vent or anus.

Acrosomal Cap. Covering over the head of a spern which is derived from the golgi body.

Acrosome. Thin-walled vesicle that forms a cap on the head of a spermatozoon. On contact with an ovum it bursts, releasing powerful lytic enzymes which cause a localized softening of the vitelline membrane, thus facilitating fertilization.

Acrospore. Store at the end of hypha.

Acrostichoid. Acrostichum.

Acrostichoid Condition. The situation found in certain ferms, *e.g* Platycerium, in which the sporangia are formed all over the undersurface of a fertile frond rather than in specialized organs or sori.

Acrostichoid Sorus. A firm sorus havin several fused sori.

Acrostichum (family Pteridaceae). Genus of ferns in which the rhizome is creeping) the leaves are pinnate, and the upper pinnae are reduced and covered below by sporangia (the 'acrostichoid' condition, which also evolved independently in other ferns). There are three species, in mangrove forests throughout the topics (A. aureum is the commonest).

Acrotretida (class Inarticulata). Order of brachiopods, consisting of two suborders, which are usually circular or semicircular tn outline. The shell is either phosphatic or punctate calcareous. The pedicle opening is restricted to the pedicle valve. The beak is marginal to subcentral in position.

Acrotretidina (order Acrotretida). Suborder of inarticulate brachiopods with phosphatic shells. Throughout life they are attached to the sea-bed by a pedicle. They first appeared in the Lower Cambrian. There are three superfamilies, seven families and about 50 genera.

Acrozone. Range zone.

ACTH (Adrenocorticotropic Hormone). A peptide which is secreted from anterior pituitary lobe (sdenohypophysis); acts on adrenal cortex for increasing the synthesis and secretion of steroids in adrenal cortex thereby resulting in conversation reaction during stress. (Also homeostasis). The corticotropic releasing hormone (CRH) from hypothalamus stimulates release of the hormone while the hormone itself is responsible for inhibiting ite release.

Actin. Globular protein of relative molecular mass 60 000 which is a major component of micro-filaments generally, and which is especially important in the myofilaments of striated muscle cell.

Actinal Surface. The oral surface, *i e.* that on which the mouth is situated in an animal that is showing radial symmetry.

Actiniaria (sea-anemones; subclass Zoantharia). Order of solitary polyps without skeletons, but with numerous tentacles and paired mesenteries. There are about 200 living genera, with 700 species.

Actiniarian. A zoantharian coelenterate of the order Actiniaria which includes the sea anemones.

Actinistia

1. Alternative name for the Crosspterygii.
2. Alternative name for a superorder of ancient, tassel-finned fish (Coelacanthimorpha) with only one living relative, Latimeria chalumnae, of E. Africa.

Actinoceratida (class Cephalopoda, subclass Naiutilidea). Order of predominantly orthoconic cephalopods that have large siphuncles. Large amounts of siphuncular and cameral deposits are present. Most are assumed to have been niktonic. It is a large order, ranging in age from Lower Ordovician to Lower Carboniferous.

Actinodromous. Describing a form of leaf venation in which three or more primary veins originate at the base of the lamina and run out towards the margin. The leaves of sycamore (Acerpseudoplatanus) are an example. In traditional terminology such venation is termed palmate or digiate.

Actinomorphic. Flowers are regularly arranged and can be cut into two similar halves in two or more directions.

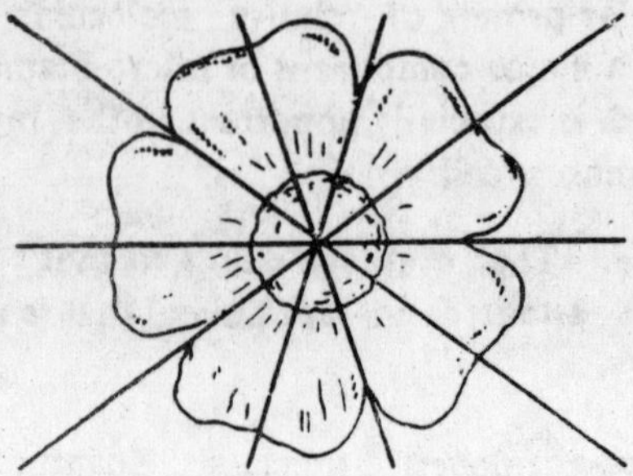

Actinomorphic flower

Actinomorphy. Radial symmetry.

Actinomycetaceae (order Actinomycetales). Family of bacteria that form pleomorphic rods or filaments that fragment readily; no mycelium is formed. Species are aerobic, facultatively anaerobic, or anaerobic. There are five genera, with many species, found chiefly in the alimentary tract in man, other mammals, birds or insects; a few species are pathogenic.

Actinomycetales. An order of bacteria containing forms that develop a branched mycelium reminiscent of fungal growth forms, though on a far smaller scale (actinomyce etc., hyphae rarely exceed 1.5 μm in diameter). Most are saprophytic though same are pathogenic.

Actinomycetes. A group of Gram-positive mostly anaerobic nonmotile bacteria. All species are fungus-like, with with filamentous cells producing reproductive spores on aerial branches similar to the spores of certain moulds. The group includes bacteria of the genera Actinomyces, some species of which cause disease in animals and man; and Streptomyces which are a source of many important antibiotics (including streptomycin).

Actinomycosis. Any disease of humans or other animals in which the causal agent is an actinomycete.

Actinophage. Virus that infects actinomycetes.

Actinoplanaceae (order Actinomycetales). Family of aerobic, mycelium-forming bacteria. Spores, which may be motile, are formed in closed sacs at the tips of hyphae held aloft from the substrate. They are saprotrophic and are found in soils rich in humus, and on dead plant or animal material. There are 10 genera, with many species.

Actinopodea (superclass Sarcodina). Class of protozoa that are typically spherical amoebae with radiating pseudodia. They are found free-living in freshwater and marine environments.

Actionopterygii (ray-finned fish). A subclass of the Osteichthyes (bony fish), comprising the ray-finned fish, which include the majority of living bony fish of sea and fresh water. The fins are composed of a membranous web of skin supported by a varying number of spines and soft rays. This subclass includes a diversity of fish types, ranging from the sturgeon and paddlefish to the eel and tuna. They appeared first during the Devonian.

Actinostele. A protostele in which the central core of xylem is star shaped or somewhat lobed as viewed in transverse section. This type of stele is exhibited in various species of Psilotum and Lycopodium and in the roots of higher plants. Depending on the number of xylem lobes, usually between two and eight, the stele is described as diarch, triarch, etc., and polyarch if the number of lobes exceeds eight.

Actinostrobus (family Cupressaceae). Genus of gymnosperms, endemic to Australia, and found as emergent tree species on heaths and sands in the south and west. A. pyramidalis (swamp cypress) is a pyramidal conifer up to 4 m in height, with dense, bright-green foliage reching to the ground. It is found in W. Australia and is now cultivated. There are three species.

Actinotrichium pl. **Actinotrichia**. A minute, horny rod, a number of which develop at the tips of fins in certain bony fishes.

Actinotroch Larva. In the phylum Phoronida, and elongated, ciliated trochophorelike larva.

Actinotus (family Umbelliferae). Genus that is thought to link the evolution of endemic Australian genera of Umbelliferae to cosmopolitan genera. Known as 'flannel flowers', most members of the genus are biennial herbs found in areas of low fertility, and in some arid regions. There are 17 species found in Australia, and one in New Zealand.

Actinula. The creeping larval stage of certain tubularian hydrozoans.

Action Potential. The transitory charge in electrical potential occurring across the membrane of a nerve or muscle fibre during the passage of a nervous impulse. The degree of change has been found to be independent of the strength of the impulse; it either takes place or does not (see all-or-none). In the absence of an impulse there is a resting potential of about-70 mv across the membrane (inside negative) due to the intracellular and extracellular media (see sodium pump). However, during the passage of an impulse the resistance of the membrane decreases and permits an inward current of sodium ions, which makes the inside less negative and thereby change the membrane potential to about+30 mv (inside positive). It is followed by an outward flow of potassium ions shortly afterwards and restores the resting potential. Local currents are following ahead of the action potential that in turn result in further action potentials. In this way a wave of activity is propagated along the membrane to give rize the impulse.

Action Spectrum. A plot of the rate of a reaction, *e.g.* the phototropic response or photosynthesis, at different wavelengths of light. The action spectrum for photosynthesis (diagram) is obtained by measuring the photosynthetic yield for a given amount of light incident upon the plant over a range of wavelengths. For green plants the action spectrum shows that chlorophyll is the pigment responsible for photosynthesis, since peak photosynthetic activity cccurs at the absorption peaks of chlorophylls *a* and *b*.

For some algae the action spectrum shows peak photosynthetic activity at the absorption peak of one of the accessory pigments. In this situation although chlorophyll is still used for transfer of light energy into the photosynthetic process, it is not the primary light-absorbing pigment.

Activation Energy (energy of activation). The energy that must be delivered to a system in order to increase the incidence within it of reactive molecules, thus initiating a reaction. It is an important feature of enzymes that they greatly lower the activation energy of many metabolic reactions.

Activator. Metal ion that functions in conjunction with either an enzyme or its substrate in order to bring about a reaction.

Active Centre (site). The part of an enzyme molecule that undergoes combination with the substrate. A relatively small number of the atoms of the enzyme molecule gets involved.

Active Immunity. Resistance to a disease acquired by an animal as the result of the production of antibodies in response to antigens produced by the disease organism whilst inside the host animal.

Active Layer. Seasonally thawed surface layer between a few centimetres and about 3 m thick, lying above the permanently frozen horizon in a periglacial environment. It may be subject to considerable expansion on freezing, especially if silt-sized particles dominate, with important engineering implications.

Active Pool. The part of a biogeochemical cycle in which the nutrient element under consideration exchanges rapidly between the biotic and abiotic components. Usually it is smaller than the reservoir pool, and is sometimes referred to as the 'exchange' or 'cycling' pool.

Active Site. Refers to the region of an enzyme molecule that undergoes combination with and acts on the substrate. It consists of amino acids which are arranged in a configuration specific to a particular substrate or type of substrate. Binding of an inhibiting compound elsewhere on the enzyme molecule may alter this configuration and thereby the efficiency of the enzyme activity.

Active Transport. Refers to the transport of molecules or ions taking place acro s a cell membrane against a concentration gradient, with the expenditure of energy. Anything which is interfering with the provision of energy will interfere with active transport. The exact mechanism is not known with certaintly but there is a carrier involved, probably a protein or lipoprotein.

Actomyosin. A fibrous compound of two proteins, actin and myosin, present in muscles. Shortening of the fibres causes muscular contraction and enables movement.

Aculeata (wasps, ants, bees; order Hymenoptera, suborder Apocrita). Division of apocritan Hymenoptera in which the ovipositor has lost its egg-laying function and is modified as a sting for the injection of venom. In hunting wasps this last function is used to paralyse, but not kill, insect or spider prey. The sting is also a means of defence, and this is its exclusive function in bees. The sting has been lost in some aculeates (*e.g.* formicine ants) and is reduced or absent in some bees (*e.g.* the Andrenidae and Meliponini). Compare Parasitics.

Aculeate

1. Prickly, pointed.

2. Applied to organisms that are armed with a sting.

3. Appliee to members of the hymenopteran division Aculeata (which have sings). The word is derived from the Latin aculeatus, meaning sting, from acuts, needle.

Aculeus

1. A prickle or spine, as that of a rose.
2. A sting or stinger, as that of a honeybee.

Acumen. The subtriangular apex of the rostrum in certain crustaceans.

Acuminate. Gradually narrowing to a point, as do certain leaves.

Acupuncture. Old system of medical trertment which involves the piercing the body at specific points in specific number and pattern with metalic needles (particularly Ag, Au, Zn, etc.) for treating diseases.

Acute. Having a sharp but not extended point, as certain types of leaf.

Acute (of disease). Applied to a disease that develops rapidly and is of short duration; symptoms tend to be severe.

Acyclic

1. Not occurring in cycles.
2. *bot. a.* Not arranged in circles or whorls. *b.* Designating a flower with parts inserted spirally on the receptacle.

Acyl Carrier Protein. (ACP). A low-molecular-weight conjugated protein. It forms a complex with the enzymes of fatty acid synthesis and binds the acyl group of the growing fatty acid moiecule, thus bringing enzymes and substrate together. The prostthetic group of ACP, 4′-phosphopantetheine, is identical to the acyl-carrying portion of coenzyme A.

Acylglycerol (glyceride). An ester of the alcohol glycerol in which a hydroxyl group is replaced by a fatty acid. Acylglycerols are one of the four main types of complex lipids. Often all three of the glycerol hydroxyl groups are esterified giving a triacylglycerol. Some diacrylglycerols contain a sugar attached to the unesterified hydroxyl group.

Adactyl. Lacking fingers or toes.

Adamkiewicz's Test. Hopkins-Cole reaction.

Adansonia (family Bombacaceae). Genus of trees, several species of which are pollinated by ants that inhabit modified spines. Baobab (A. digtata) is famous for its hugely swollen trunk, commonly 15 m girth, and reaching 35 m. Other species swell, but less so. Baobab provides human and animal food and medicines. There are ten species, occurring in the seasonal tropics of Africa, Madagascar, India, and Australia.

Adapidae (order Primates, suborder Prosimii, infra-order Lemuriformes). Extinct family of lemur-like animals. Most were small, but a few had heads 10 cm or more long. The brain case was small but with temporal crests, smooth cerebral hemispheres, and the tympanic ring was included in the bulla. Dentition was full and the incisors were not procumbent, but in Adapis species the canines were incisiform, the molars resembling those of some modern lemurs. The skeleton was adapted for grasping, leaping and perching. Old World adapids (*e.g.*, Adapis and Pronycticebus) probably resembled lemurs and lorises, to which they may be ancestral. They were distributed in Europe and N. America in the Eocene.

Adaptation

1. A modification to an organism, or a feature of an organism, that makes it better fitted for a particular environment. Adaptations may either be acquired

during the life of the individual or, if governed by the genotype, be inherited.

2. (in physiology) The alteration in the degree of sensitivity (either an increase or a decrease) of a sense organ to suit conditions more extreme than normally encounted. An example is the adjustment of the eye to vision in very bright or very dim light.

Adaptive Breakthrough. Evolutionary change by the acquisition of a distinctive adaptation that permits a population or taxon to move from one adaptive zone to another. At the most extreme such moves might be from water to land, or from land to air.

Adaptive Conzergence. Refers to the superficial similarity between different species of animals due to similarity of habits, or sometimes due to the fact that both are exhibiting protective resemblance to some other species, *e.g.*, several flies, beetles, spiders etc. show a protective resemblance to ants.

Adaptive Enzyme. An enzyme formed by an organism especially by bacteria in response to growing on an substrate, *i.e.*, the enzyme is not normally present, but develops soon after growth on the new substrate has started, so that it can be used.

Adaptive Orientation. A habit which is shown by many insects as well as some birds and animals, of standing or resting in such an attitude as to make the best possible use of their protective coloration.

Adaptive Pathway. Series of small adaptive steps, rather than a single large one, which leads from one adaptive zone across an enviromental and adaptive threshold into another adaptive zone. In effect, small changes accumulate so that the organism is virtuolly pre-adapted to enter the new zone.

Adaptive Peaks and Valleys. Symbolic contour map showing the Darwinian fitness or adaptive value of genotypic combinations will usually display adaptive peaks and valleys : these occur at points at which the fitness is respectively strong or weak. The distribution of the population of a given species will therefore coincide with that of the adaptive peaks.

Adaptive Radiation (divergent evolution). The evolution from one species of animals or plants of a number of different forms. As the original population increases in size it spreads out from its centre of origin to exploit new habitats and food sources. In time this results in a number of populations each adapted to its particular habitat : eventually these populations will differ from each other sufficiently to become new species. A good example of this process is the evolution of the Australian marsupials into species adapted as carnivores, herbivores, burrowers, fliers, etc. On a smaller scale, the adaptive radiation of the Galapagos finches provided Darwin with crucial evidence for his theory of evolution.

Adaptive Type. Population or taxon with distinctive adaptive attributes, expressed as a particular morphological theme, characteristic of a particular habitat or mode of life. In evolutionary terms, the appearance of a new adaptive type is frequently followed by radiations that yield variants that partion the environment and exploit it more effectively.

Adaptive Value (Drawinian fitness, selective value). The balance of genetic advantages and disadvantages that determines the ability of an individual organism (or genotype) to survive and reproduce in a given environment. The 'fittest' is the individual (or genotype) that produces the largest number of offspring that later reach reproductive maturity.

Adaptive Zone. Comprises a taxon and the associated environmental regime or regimes, habitat, or niche.

Adaptor Compound. A carrier compound, *e.g.*, RNA, which is responsible for transferring a substance to the site of reaction .

Adaxial. Describing the side of lateral organs facing towards the stem or main axis, *i e*, the upper surface. Compare abaxial.

Adcrustation. The deposition of a substance on a cell-wall, instead of being laid down between the existing molecules. This latter process is termed as encrusting.

Adder. Also called the common viper, this poisonous snake is found throughout Europe except in the extreme south and Ireland. It grows to about 65 cm long and is characterized by the dark zig-zag markings along its grey or

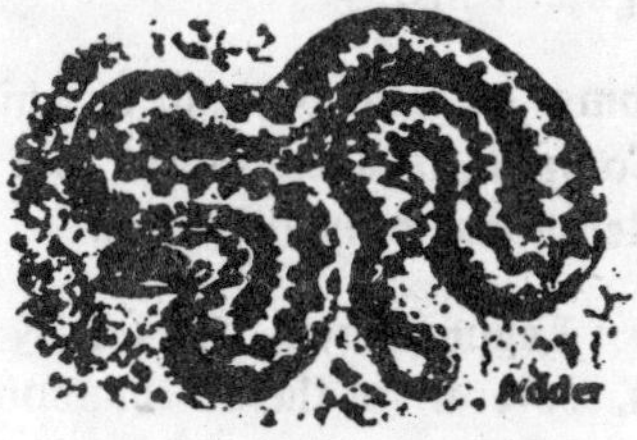

Adder

brown back. It also has vertical pupils in its eyes, unlike all other NW European snakes, which have round pupils. Adders give birth to live young enclosed in a thin membrane.

Additive Fixative. A fixative that undergoes chemical combination with the protein of the material to be preserved.

Adduct. To draw toward a median plane or axis.

Adductor

(1) A muscle is pulling a limb towards the main axis of the body.

(2) A muscle which is pulling two parts together, as for example those which close a bivalve shell.

Adeciduate. Not dediduous.

Adeleina (order Eucoccidia). Suborder of parasitic protozoa characterized by the occurrence of syzygy. The parasites develop in epithelial cells and cells of the circulatory systems in vertebrate and invertebrate hosts. There are several families.

Adelie Penguin. A small penguin standing about 45 cm high which lives on Antarctica and neighbouring islands. It has simple black and white markings and a comic, waddling gait. To move faster, these penguins toboggan over the ice on their bellies, using their feet and flippers to propel themselves.

Adelphogamy. Union between a vegetative mother-cell and one of its daughter cells.

Adelphous. Describing an androecium in which the filaments are fused. Compare syngenesious. See also, mona-delphous, diadelphous, polyadelphous.

Adenanhera (family Leguminosae, subfamily Mimosoideae). Genus of trees, several of them common as shade trees. A pavonina has hard, scarlet seeds, used for centuries in India and Sri Lanka as units in one of the oldest weight systems known, for weighing gold, etc. There are 12 species, occurring from S. China to the Pacific islands.

Adenanthos (family Proteaceae). Genus of shrubs with single flowers, often bright red. Height varies from that of a 3 m bush to prostrate forms They are similar to Grevillea. There are 32 species, endemic to W. Australia

Adenine. A nitrogenous base, more correctly described as 6-aminopurine, derived from amino acids and sugars and found in all living organisms. Adenine is a constituent of *DNA, *RNA, *ATP, *NAD, and *FAD. Many cyto-kinins are derivatives of adenine and adenine itself shows cytokinin activity.

Adenohypophysis. The portion of the pituitary gland which develops from Rathke's pouch. It comprises the pars tuberalis and pars distalis of the anterior lobe, and the pars intermedia.

Adenoids. A mass of hypertrophted pharyngeal tonsil sometimes blocking the posterior nares.

Adenosine. A combination of adenine with D-ribose, a pentose sugar. The two molecules are linked by a β-glycosidic bond and together form a nucleoside. Adenine may alternatively combine with 2-deoxy-D-ribose, in which case the nucleoside is called deoxyadenosine. The latter is the form found in DNA, while adenosine is found in RNA, ATP, ADP, and AMP. Certain derivatives of adenosine (*e g.* isopentenyl adenosine or IPA) are cytokinins.

Adenosine Diphosphate (ADP). High-energy phosphoric ester, or nucleotide, of the nucleoside adenosine. It can undergo hydrolysis to adenosine monophosphate and inorganic phosphate, the reaction releasing 34 kJ/mol of energy at pH 7.

Adenosine Monophosphate. A compound composed of adenine, d-ribose, and phosphoric acid, of importance in the release of energy for cellular activity. Also called AMP, adenylic acid, adenine ribotide.

Adenosine Triphosphate. A compound present in all cells which provides energy derived from food or sunlight for energy-expending processes, as protein synthesis, muscle contraction, impulse conduction and glandular secretion. Upon hydrolysis, it yields adenosione diphosphate (ADP) and a simple phosphate—also called ATP.

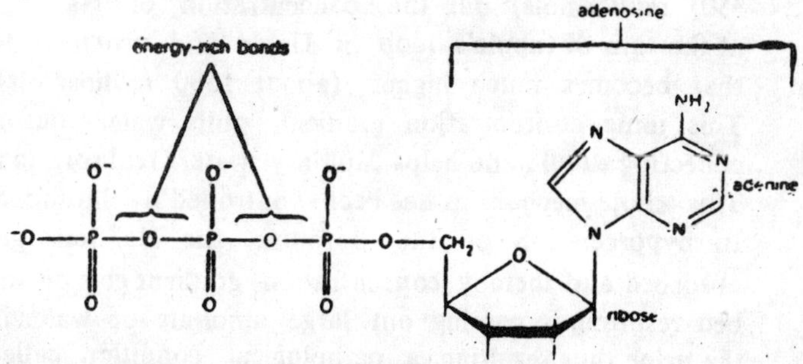

A.T.P.

Adenotrophic Viviparity. Used for describing the phenomenon in certain insects (*e.g.* Glossina) in which the larvac after hatching are retained in the 'uterus' of the parent, develop by feeding on uterine secretions, moult twice and are ultimately deposited as mature larvae are almost ready to pupate.

Adenovirus. One of a group of DNA-containing viruses found in rodents, fowl, cattle, monkeys, and man. In man they produce acute respiratory-tract infections with symptoms resembling those of the common cold. They are also implicated in the formation of tumours.

Adenylate Cyclase. Refers to an enzyme which is associated with hormone action at its target cell catalysing the hydrolysis of ATP (adenosine triphosphate) to c-AMP when it gets allosterically activated due to conformational changes taking place in receptor molecule present in target cell membrane which is brought about by interaction of receptor molecule. The compound has been closely associated with receptor molecules but faces towards cytoplasm.

ADH (Antidiuretic Hormone). Refers to hormone which is secreted by pos terior pituitary lobe. It is involved in reabsorption of water from glomerular filtrate and thereby controls urine formation (diuresis), In kidney tubule Na^+ concentrarion around the proximal and distal convoluted tubules are equivalent to that of most of body (300-350) milliosmols) but the concentration of Na^+ ions at the turn of tubule's loop of Henle and proximal to that becomes much higher (about 1500 milliosmols). This ionic concentration gradient pulls water out of collecting tubule and helps capillary water reabsorption. This whole mechanism has been controlled by hormone. In hyposecretion of the hormone the Na^+ ions get absorbed and thereby concentration gradient gets disturbed resulting in passing out large amounts of water in the urine thus resulting a pathological condition called Diabetes insipidns.

Adherens Junction. Cell junction commonly observed in epithelial cells, *e.g.* those lining the intestine and those in cradiac muscle cells. At these junctions the plasma membranes of the neighbouring cells are separated by a space of 15—25 nm which is filled with a filamentous material. Beneath this, anchoring the junction to the cell cytoplasm, are loosely structured mats of fibres, 7 nm in diameter, thought to be actin filaments. It is thought that these junctions may provide mobility in the regions of the cells in which they are located.

Adhesion. The union of dissimilar organs.

Adhesive. Sticky; tending to adhere, as adhesive cells, glands, and pads of various invertebrates.

Adiantaceae. Family of ferns in which sori are borne on the undersides of the veins of the pinnules, without indusia. Their distribution is cosmopolitan but they occur most abundantly in the moist tropics. There are about 60 genera.

Adiantum (maidenhair ferns; family Adiantaceae). Genus of ferns in which all the fronds are alike, usually with black, glossy petioles and wide, fan-shaped leaflets. Rhizomes are scaly. Sori are borne near the vein ends on the reflexed tips of the leaflets. There are about 200 species. Most occur in the moist tropics and warm-temperate regions, as far as the southern British Isles.

Adient. Turning toward or approaching a source of stimula.

Adipose. Fatty; fatlike.

Adipose Eyelid. The thickened yet transparent skin overlying the eyes of some animals, including some fish, such as certain herring and mullet species. Apart from a small central aperture, it covers most of the eye, giving the fish a 'bespectacled' appearance.

Adipose Fin. In some fish, a type of second dorsal fin in the form of a small flap of skin tissue without supporting rays. Largely composed of fatty tissue covered with skin, it is found amongst members of the salmon, catfish, and a few other families.

Adipose Tissue. A body tissue comprising cells containing fat and oil. It is found chiefly below the skin (subcutaneous tissue) and around major organs (such as the kidneys and heart), acting as an energy reserve and also providing insulation and protection.

Adiposis. Excessive accumulation of fat; obesity.

Aditus. *anat.* An entryway.

Adjuvant. An agent which helps, enhances, or increases the effectiveness of another substance, as a medicine or medical treatment.

Adnate. Describing unlike organs that are joined together, such as stamens fused with the petals. Compare connate.

Adnation. *bot.* The fusion of members of different floral circles.

Adnexed

1. Applied to the gill of an agaric that is attached to the stipe by only a part of its width.

2. Applied to analogous loose attachments of two organs in higher plants. Compare adnate.

Adrenal Glands. A pair of glands present above each kidney and produce the hormone adrenaline. Secretion is controlled by the nervous system.

Adnnate. Used for describing organs which are joined together *e.g.*, stamens fused with the petals. Compare conate.

Adolescariae. The cercariae and metacercariae of flukes. Compare marita, parthenita.

Adolescence. Refers to the period of life-cycle of an organism during which the quiscent repparatus develaps and becomes fully operational. In man it starts at the age of twelve years and in women at eleven years of age.

Adoral. Near or toward the mouth. Compare aboral.

Adoxaceae. Family of one monotypic genus of obscure affinities, Adoxa (moschatel), of almost circumpolar distribution in the northern-hemi sphere, cool-temperate woodlands. A herb 5—10 cm tall, with ternate leaves and a terminal inflorescence of four outward- and one upward-facing flowers which are five- or (in the terminal flower) four-lobed and light green; the ovary is inferior. The fruit is a drupe.

ADP (adenosine diphosphate). A diphosphorylated nucleoside with the same structure as adenosine monophosphate but with a second phosphate linked through a high-energy phosphate bond to the first. In photophosphorylation ADP combines with another phosphate molccule to form ATP. This conversion is a major mechanism for the conservation of the light energy absorbed in photosynthesis. In some plants ADP acts as a sugar carrier, forming such molecules as ADP-glucose and ADP-sucrose. These ADP-sugars can then be oxidized to give energy or used in the synthesis of starch and ascorbic acid.

Adrectal. Near or close to the rectum.

Adrenal Cortex. Refers to the outer part of adrenal, gland; divided into three zones of variable distinctness—outer zona glomerulosa having whorls of cells which become continuous with columns of cells of middle zone called zona fasciculata (these columns are separated by venous sinuses), innermost zona reticularis where all the columns are interlaced in network. The cells have abundant lipids which are present especially more in outer portion of zona fasciculata. It secretes glucocorticoids which are having widespread effects on metabolism of carbohydrates and proteins, mineralocorticoids which play essential role in maintenance of sodium balance and extracellular fluid volume and sex hormones which tend to exert minor effects on the reproductive functions.

Adrenal Gland. In vertebrates, a gland of internal secretion located near the kidney. In mammals, it consists of an outer cortex of mesodermal origin and a medulla of ectodermal origin. The cortex secretes a number of steroid hormones (sex hormones, glucocorticoids, mineralocorticoids); the medulla secretes epinephrine and norepinephrine; also called suprarenal gland.

Adrenaline (epinephrine). A hormone, produced by the medulla of the adrenal glands, that increases heart activity, improves the power and prolongs the action of muscles, and increases the rate and depth of breathing to prepare the body for 'fright, flight, or fight'. At the same time it inhibits digestion and excretion. Similar effects are produced by stimulation of the sympathetic nervous system. Adrenaline can be administered by injection to relieve bronchial asthma and reduce blood loss during surgery by constricting blood vessels.

Adrenal Medulla. Refers to the portion of adrenal gland; consists of interlacting cords of densely intervated granule-containing cells which about on venous sinuses. Morphologically there are two types of cells-epinephrine secreting cells (having dense granules) and nor-epinephrine secreting cells (in which the smaller dense granules do not fill the vesicles in which they are present). Paraganglia—small groups of cells resembling those of medulla, occur near the thoracic and abdominal sympathetic ganglia. Medulla secretes epinephrine (adrenaline) and nor-epinephrine (nor-adrenaline) which are collectively known as catecholamines.

Adrenergic A term for describing nerves which liberate a substance similar to adrenalin at their endings, *e.g.* many of the sympathetic nerves. Nerves of this type usually possess an action antagonistic to the other type which secrete acetylcholine at their endings. These are termed as cholinergic. An example of this antagonism is, that adrenergic nerves increase the beating of the heart and cholinergic nerves show it down.

Adrenergic Receptor. Refers to the membrane receptor molecules which are specific to interact with catecholamines; type of nerve fibre which releases adrenaline like substance from axon terminal when nerve impulse transmission is taking place across the synapses. It is possible to distinguish two types of adrenergic receptors (α-adrenergic receptors and β-adrenergic receptors) by use of drugs that activate one or the other receptor types, eliciting responses which are characteristic of one or the other type of catecholamine.

α-Adrenergic Receptors. Refer to the adrenergic receptors to which adrenaline (epinephrine) binds most effectively.

β-Adrenergic Receptors. Refer to the adrenergic receptors to which adrenaline (epinephrine) binds most effectively.

Adrenocortical. Pertaining to the adrenal cortex.

Adrenocorticotrophic Hormone. A hormone secreted by the anterior lobe of the hypophysis which stimulates secretion by the adrenal cortex, especially the secretion of glucocorticoids—also called ACTH.

Adrenotrophic Hormone. A hormone which is produced by the anterior pituitary lobe, stimulates activity of the adrenal glands.

Adsorption. The formation of a layer of solid, liquid, or gas on the surface of a solid or, less frequently, of a liquid. There are two types depending on the nature of the forces involved. In chemisorption a single layer of molecules, atoms, or ions is attached to the adsorbent surface by chemical bonds. In physisorption adsorbed molecules are held by the weaker physical forces. The property is utilized in adsorption chromatography.

Adult. A fully grown, sexually mature individual.

Aduncate. Hooked.

Adventitia. Said of an outer layer which is surrounding some structure, *e.g.*, dense fibrous connective tissue layer which is surrounding alimentary canal in neck and thorax region where serosa is not present.

Adventitious

1. Additional; accidental; foreign; casual.
2. Occurring in unusual or abnormal positions, as adventitious roots or buds.

Adventitious Roots. Roots which develop from the stem or leaf, rather than as branches from the main root.

Adventive. Net being or having become natural to a place, said of plants taken from other places.

Adventive Embryo (Adventitious embryogenesis). The formation of an embryo in some position other than within an embryo sac.

Adventive Embryony (adventitious embryogenesis). The formation of an embryo in some position other than within an embryo sac. It is a form of apomixis. Adventive embryos may develop from somatic cells of the nucellus or chalaza (apospory) and may occur together with a normal zygotic embryo. One seed can contain several embryos, as is often the case in Citrus seeds. Alternatively an embryo may develop from the unreduced egg cell (diplospory).

Advertisement. Form of display in which an individual makes itself as conspicuous as possible. It is used most commonly by male animals holding a territory, in order to ward off rivals and to attract females.

Aeciospore (aecidiospore). A dikaryotic spore formed by certain rust fungi in a small cup-shaped sorus, the aecium or aecidium. The aeciospores of Puccinia graminis are formed on the lower surfaces of barberry leaves in clusters of aecia, this stage of the life cycle commonly being called the cluster-cup stage. The aeciospores cannot reinfect barberry but will germinate on certain grass or cereal leaves where the subsequent infection shortly gives rise to uredospores.

Aecium (aecidiosorus). Structure formed by certain rust fungi (order Uredinales) in the tissues of a host plant. A mature aecium is often cup-shaped, with chains of spores (aeciospores) being produced from the base of the cup; in some rust fungi the aecium is a more diffuse structure.

Aedeagus Intromittent organ or penis of males of most insect groups, which is of great diagnostic value in many groups. The inner wall of the aedeagus is a continuation of the ejaculatory duct and is usually called the endophallus It may be inflatable and everted during copulation. It has a gonopore at its base and opens at its apex through a phallotreme, the whole structure being surrounded by a sclerotized cuticle. The endophallus may be armed with backward-directed teeth or spines which gain purchase on the veginal walls of the female.

Aedes Aegypti (yellow-fever mosquito; order Diptera, family Culicidae) Cosmopolitan mosquito of the tropics and subtropics, where the females transmit the virus of yellow fever when they bite. It is a coastal and riverine species, and rarely breeds far from human habitations. The eggs are laid in small areas of stagnant water in tree holes, old tins, and bottles, or in broken coconut shells. The larvae are colourless, with a black respiratory siphon. The adult bears a characteristic, white, lyre-shaped mark on the dorsum of the thorax.

Aegithalidae (long-tailed tits, bushtits; order Passeriformes). Family of small titmice with medium to long tails. The nest is intricate and domed, with a small entrance, built in trees and bushes. Aegithalos (long-tailed tits) is a Eurasian genus, Psaltriparus (bushtit) is an American genus, and Psaltria (pygmy tit) is confined to Java.

Aegotheles (owlet-nightjars; family Aegothelidae). Genus of brown-grey, secretive, nocturnal birds similar to Podargidae (frogmouths) and Caprimulgidae (nightjars). They are insectivorous, with small, broad bills. They prossess preen glands. They nest is holes in trees or the ground. There are eight species, found in Australasia, (but not in New Zealand).

Aequorin. A type of an animal protein extracted from the jellyfish Aequorea. When it is combined with calcium ions, it emits a photon of bluish-green colour. Discovered in 1963. The protein finds use in the measurement of the Ca^{2+} in various tissues and body regions photocolorimetrically.

Aerating Root. A root of some plants which grows in soft mud They have aerenchyma, and stand above the surface of the ground acting as ventilating organs.

Aeration

1. Exposure to air.
2. Impregnation of a fluid with air or oxygen.
3. Oxygenation of the blood in the lungs.

Aeration Tissue. Aerenchyma.

Aerator. A device for supplying air or oxygen.

Aerenchyma. Tissue with numerous large intercellular spaces. Aerenchyma is common in the cortex of the roots and stems of many aquatic plants, where it facilitates oxygenation of the roots and may increase buoyancy.

Aerial

1. Of or pertaining to air.
2. Living in or frequenting the air.
3. Occurring in, produced by, or found in air.

Aerial Plant. Epiphyte.

Aerial Roots. Roots which develop partly or entirely above the ground, as the roots of orchids.

Aerobe. An organism capable of living only in the presence of air or free oxygen, the oxygen being needed for aerobic respiration. All plants, except certain bacteria and fungi, are aerobes. In contrast to most animals they do not die instantly when deprived of oxygen but can continue to respire anaerobically for a time. Compare anaerobe.

Aerobiont. Same as aerobe.

Aerobic Respiration. A type of respiration in which foodstuffs (usually carbohydrates) are completely oxidized to carbon dioxide and water, with the release of chemical energy, in a process requiring atmospheric oxygen. The reaction can be summarized by the equation :

$$C_6H_{12}O_6 + 6O_2 \rightarrow 6CO_2 + 6H_2O + \text{energy}$$

Most organisms have aerobic respiration (*i.e.* they are aerobes); exceptions include certain bacteria and yeasts. Compare anaerobic respiration.

Aerobiology. The study of organisms of their products (spores, pollen) present in the atmosphere.

Aerobiosis. Life necessitating the presence of air or free oxygen.

Aerodynamic Method. Method for monitoring rates of primary productivity; it is particularly useful in studies of woodland or forest communities. The carbon-dioxide concentration from ground level to above canopy height is measured at regular intervals, or continuously, by sensors attached to a mast. Differences in carbon-dioxide concentration reflect differences in rates of photosynthesis and respiration and hence in productivity. The chief advantages of this approach are that it is non-destructive and causes minimal disturbance to the environment of the vegetation being studied. Other non-destructive methods usually enclose the vegetation (*e.g.* in polythene or plexiglass chambers), which can make results unrepresentative of normal field conditions. Compare harvest methods; chlorophyll method; and gas-exchange method.

Aerolal. Used for cracks which get developed in the surface of a crustaceous lichen due to growth strains. These cracks help in aeration.

Aeropalynology. The study of pollen grains and spores in the atmosphere. This is important for allergy studies (*e.g.* of hay fever), and for understanding the spread of diseases in man, other animals, and agricultural plants; and recently for palaeoecologists attempting to improve the interpretation of microfossil data.

Aerophore. Pneumatophore.

Aerophyte. A plant which thrives in air, securing its water from rain or dew, as an orchid.

Aeroplankton. Collectively, the microorganisms present in air.

Aerotaxis. A form of chemotaxis in which the cell or organism moves in response to oxygen.

Aerotaxis (aerotatic movement). A taxis in response to an oxygen concentration gradient. For example, motile aerobic bacteria are positively aerotactic, while motile obligate anaerobic bacteria are negatively aerotactic.

Aerotropism. A form of chemotropism in which oxygen is the orientating factor.

Aesculus (horse chestnut, buckeye; family Hippocastanaceae). Genus of deciduous trees with large, sticky buds, opposite palmate leaves, and showy terminal panicles of white or red flowers. There are about 25 species, found in deciduous forests of SE Europe, E. Asia, and N. America. They are much cultivated.

Aeshnidae (dragonflies; order Odonata, suborder Anisoptera). Family of ubiquitous, large, swiftflying, hawker dragonflies which are strikingly marked with blue or green. They have similar triangles in the fore- and hindwings, a well defined but small and loop in the hindwings, the eyes broadly contiguous across the top of the head, the ocelli in a triangle around a raised tubercle, and the ovipositor complete and not projecting beyond the long anal appendages. The nymphs are unmistakable, having large eyes and slender bodies. More than 430 extant species have been described, and their distribution is cosmopolitan. Aeshna cyanea (blue aeshna, southern hawker) is a large, inquisitive dragonfly which is often encountered along tracks and hedgerows in rural areas, and may occur far from water, in towns and gardens. It is the commonest of the large hawker dragonflies of Europe. In the male there is a well defined T. shaped blank spot on the front of the head, with the sides of the thorax having broad, apple-green stripes, and the abdomen being spotted with green and blue. The female has similar green thoracic stripes and the abdomen is patterned with yellow and green on a brown background. The terminal segments contain blue in the male and green in the female. Both sexes have two broad, green, antehumeral stripes. The species occurs throughout Eurasia and N. Africa.

Aesthetasc. An olfactory hair on the antennule of certain crustaceans.

Aesthete. A sense organ, especially in an invertebrate, as those present in chitons and certain arthropods.

Aestival. In the early summer. The term is used with reference to the six-part division of the year used by ecologists, especially in relation to the study of terrestrial and freshwater communities. Compare autumnal; hibernal; prevernal; serotinal; and vernal.

Aestivation

1. (in zoology). A state of inactivity occurring in some animals, notably lungfish, during prolonged periods of drought or heat. Feeding, respiration, movement, and other bodily activities are considerably slowed down.

2. (in botany). The arrangement of the parts of a flower bud, especially of the sepals and petals.

Aethalium. A sessile, rounded fruiting body in certain slime molds formed by the heaping up of the plasmodium into a rounded mass.

Aextoxicon (family Aextoxicaceae). Genus of one species, a large tree of the warm temperate rain forests of Chile. The leaves are (sub)opposite, and lepidote. Flowers are borne in axillary racemes, and are unisexual, and pentamerous. The fruit is a drupe. Probably they are related to the Monimiaceae; and formerly they were placed incorrectly in the Euphorbiaceae.

Afferent. Carrying (nerve impulses, blood, etc). from the outer regions of a body or organ towards its centre. The term is usually applied to types of nerve fibres or blood vessels. Compare efferent.

Affinity Index. Measure of the relative similarity of composition of two samples, *e.g.* $A=c/\sqrt{(a+b)}$ where a and b are the numbers of species in one sample but not in the other and *vice versa* respectively, and c is the number of species common to both. The reciprocal, $\sqrt{(a+b)}/c$, indicates the ecological and phytosociological distance (D) between samples.

Afforestation

1. The establishment of woodland by natural succession or by the planting of trees on land where they did not grow formerly.

2. Historically, the act of defining an area of land that henceforth would become subject to forest laws.

Aflatoxin. Any of four related toxic compounds produced by the mould Aspergillus flavus. Aflatoxins bind to DNA and prevent replication and transcription. They can cause acute liver damage and cancers : man may be poisoned by eating stored peanuts and cereals contaminated with the mould.

Afoliate. Lacking leaves : Aphyllous.

African Violet. Tropical African flower often grown as a house plant. It has furry heart-shaped leaves, and is not a true violet.

Afterbirth. The placenta, umbifical cord, and extraembryonic membranes, which are expelled from the womb after a mammalian fetus is born. In most nonhuman mammals the afterbirth, which contains nutrients and might otherwise attract predators, is eaten by the female.

After Discharge

1. In receptors, the continuation of the discharge of impulses after stimulation has ceased.
2. In reflexes, the continuation of the motor response after stimulation of the receptor has ceased.

Afterimage. Persistence of the sensation of light after cessation of the light stimulus. In colour vision, the afterimage may be positive (of the same colour) or negative (of the complementary color).

After-ripening. The processes that must be undergone by certain seeds after harvest before germination can take place. Although the embryo appears fully mature the seed remains dormant until the after-ripening period is completed. It has been suggested that after-ripening involves the breakdown of some growth inhibitor or the formation of a growthpromoting substance. However, since after-ripening processes often occur in dry seeds, in which metabolic activity is very low, it would appear that the changes are not necessarily metabolic in nature. An after-ripening requirement is exhibited by many cereals and grasses and prevents the seed from germinating in the ear under moist conditions.

Aftershaft. An accessory feather attached at the junction of the rachis and quill.

Agama. A group of lizards related to the iguanas. The common agama of West Africa is about 30 cm long and has a variety of colours ranging from brown to vivid orange and blue. Other agamas include the Moloch and Frilled Lizard of Australia.

Agamandroecious. Male and neuter flowers present in the same inflorescence.

Agamete. A single cell other than a gamete which is capable of developing into a new organism.

Agamic. Asexual; reproduction without union of sex cells; agamous.

Agamic Generation (asexual generation). In cynipid, heterogynous wasps, parthenogenetically reproducing females that are genetically diploid. The unfertilized eggs laid may be male (having undergone meiosis, or reduction-division) or female (not having undergone meiosis). Agamic females usually emerge from stout galls designed to overwinter or to survive a period of scarce resources.

Agamidae (agamids; order Squamata, suborder Sauria (Lacertilia). Family of lizards that closely resemble iguanas, but that have acrodont teeth. Scales are keeled and often spiny. Typically they have a large head, long limbs, and a long tail with no autotomy. Dorsal crests, throat sacs, and colour change are common. The family includes bipedal and gliding forms. There are 300 species, all found in the Old World.

Agamogenesis. Asexual Reproduction.

Agamogony. Asexual reproduction as occurs in protozoa : Schizogony.

Agamogynoecious. Female and neuter flowers present in the same inflorescence.

Agamohermaphrodite. Hermaphrodite and neuter flowers pirsent in the same inflorescence.

Agmospermy. The formation of seeds without the fusion of gametes.

Agamotropic. Used for a flower which does not shut after having once opened.

Agnatha. The class having the earliest and most primitive vertebrates that are characterized by the absence of jaws. The only living order is the Cyclostomata, which is including lampreys (*e.g.* Petromyzon) and hagfish (*e.g.* Myxine).

Agnathostomata. Animal without jaws in mouth *e.g.* cyclostomes.

Agaonid, (fig wasps; suborder Apocrita, 'super-family Chalcidoidea). Family of small or minute, sexually dimorphical black, chalcidoid wasps which live within the flowers on fig trees and help pollinate them, the males, rarely emerging from the fruits. The head of the winged female is long and somewhat oblong with a deep, median, longitudinal groove; males are nearly always wingless. The front and middle legs are stout, and the front tibiae have no spurs. The hind legs are slender. The males have thick antennae, with 3—9 segments. Mating occurs within flowers and there is often fighting between males.

Agar. A mucilaginous carbohydrate that is used in the form of a gel as a microbiological supporting medium. It is obtained from certain red algae, notably Gelidium species. The gel is produced by dissolving agar crystals in boiling water and then allowing the mixture to cool. It may be sterilized by placing in an autoclave. A nutrient agar gel has various nutrients adeed to it. Agar is widely used because it is resistant to attack by almost all micro-organisms, and, as it does not melt until 100°C, incubation of the selected culture can take place at high temperatures. When agar is used in a Petri dish it is known as an agar plate.

Agaric. The agarics are a large group of fleshy fungi with gills. They include many familiar fungi such as the field mushroom, the chanterelle, the red and white spotted fly agaric and the notorious death cap.

Agaricaceae (order Agaricales). Family of fungi in which the spores in masses appear dark brown or black. The stipe breaks cleanly from the pileus and typically bears a ring. The gills are pinkish when young, becoming dark brown to black as they mature. There are three genera, including the edible mushroom. They are saprotrophic, and are found on the ground in grassland, woods, etc.

Agaricales. An order of the Hymenomycetes containing the mushrooms and toadstools. The 3250 species and 220 genera are divided into two families : the Agaricaceae, in which the hymenium is borne on the surface of gills; and the Boletaceae, in which the hymenium lines fleshy pores. Both families contain many aerial root A root that arises above soil level.

Agaric, Fly. A colourful and poisonous toadstool commonly found under birch and pine trees. The cap is bright red with white 'warts, although these are often missing in older specimens. With age the cap becomes saucer-shaped and may fade to orange.

Agaricus (family Agaricaceae) Genus of fungi which includes the cultivated mushroom, A. bisporus, as well as several edible wild species; there are also some poisonous species. They are found on the ground in fields, woods, etc., and are cosmopolitan in distribution.

Agastric. Lacking a digestive tract or cavity.

Agathis (kauri, kauri pines, Manila copal; family Araucariaceae) Genus of large, coniferous trees with broad, elliptical leaves; valuable, pale, light, timber; and resin (Manila copal) in the bark. There are 13 species, found in tropical to warm-temperate forests. They are native from Malesia to Queensland and New Zealand.

Agave. A group of desert-living American plants with rosettes of thick spiky leaves up to 3 metres long. Dense clusters to pale flowers are carried on tall central stalks. Some species take many years to flower and are called century plants. Sisal and other fibres are obtained from the leaves.

Age

1. The time from birth to the present of any living individual.
2. A particular period of life or development, as old age, the Age of Fishes, the Stone Age.

Age-and-area Hypothesis. J. C. Willis suggested in 1922 that, all other things being equal, the area occupied by a taxon is directly proportional to the age of that taxon. Thus in a polytypic genus, the species with the smallest area of distribution would be the youngest in the genus. However, things hardly ever are equal, and Willis's ideas have never gained acceptance as a law or rule.

Age Determination (in fish). The age of fish with a bony skeleton and scales can be determined by studying the number of annual rings found in certain parts of the fish. Although those in the scales are most frequently used, other structures, such as otoliths, finrays, dontary bones, and vertebrae are also analysed for this purpose.

Agelenidae (sheet-web spiders; order Araneac, suborder Araneomorphae). Family of spiders that construct sheet webs with lateral, tubular retreats, and that have a somewhat elongate appearance, with a flattened thorax. The sternum is wide, heart shaped, and may project between the fourth coxae. The chelicerae are usually very convex, and nearly vertical, the outer margin with three teeth, the inner with 2—8 teeth. The carapace is oval narrowed, and anteriad; and has a longitudinal depression (sometimes replaced by a dark line), with patterns of dots and bars. The head is raised and narrow, and points forward, with the eyes, usually uniform in size, arranged in two short rows. The legs are thin and long, are covered thickly with long hairs, and are spiny, especially on the third and fourth pairs. The anterior spinnerets are cylindrical, and more or less separated; the median spinnerets are similar or smaller; and the posterior spinnerets are usually two-segmented; and long. The tracheal spiracle is close to the spinnerets. All species in this family, which includes the common grass spiders of the genus Agelena, and Tegenaria domestica (house spider), catch prey on the upper surface of the sheet web. Argyroneta (water spider) is usually placed in this family, but sometimes in a family, of its own. Many species spend a long time together as couples. Distribution is worldwide.

Agenesis or Agenesia

1. A condition in which a structure fails to develop or development is incomplete; aplasia.
2. Inability to produce offspring.

Ageusia. Loss or important of the sense of taste.

Agger. *anat.* A projection or prominence.

Agglomerate

1. To group or gather together into a ball or mass as in certain protozoans.
2. Gathered together into a dense cluster, as in the heads of certain flowers.

Agglomerative Method. System of hierarchical classification that proceeds by grouping together the most similar individuals, and subsequently groups, into progressively larger and more heterogeneous units. At each stage the groups or individuals linked are those giving the minimum increase in group heterogeneity.

Agglutinate. To cluster and adhere together; to collect in masses.

Agglutination. The clumping togethsr by antibodies of microscopic foreign particles (antigens), such as red blood cells or bacteria, so that they form a visible pellet-like precipitate. Agglutination is a specific reaction, *i.e.* a particular antigen will only clump in the presence of its specific antibody; it therefore provides a means of identifying unknown bacteria and determining blood group. When red cells of incompatible blood groups (*e g.* group A and group B—ABO system) are mixed together agglutination of the cells occurs (haemagglutination).

Agglutinin. An antibody present in normal or immune serum which brings about agglutination of its antigen.

Agglutinogen. A substance which when injected into an animal induces the development of a specific agglutinin.

Aggregate. Clustered together to form a dense mass or head, as an aggregate fruit.

Aggregate Fruit. A fruitlike structure that hss developed from the carpels of a single flower and is composed of a number of separate fruits. It may be an aggregation of achenes (as in Anemone), berries (as in Actaea), drupelets (an in Rubus), follicles (as in Delphinium), or samaras (as in Liriodendron). The term etaerio is often used to mean any aggregate fruit but its use is sometimes restricted to a collection of drupelets.

Aglaia (family Meliaceae). Genus of plants, most of which are medium-sized trees. They have twigs with stellate hairs, and leaves that are large and pinnate (or, rarely, simple). The flowers are tiny, brone in panicles, and often fragrant. The fruit is one or few-seeded, the pericarp leathery and dehiscent, the read large, with a fleshy jacket. A few members of this genus are timber species (including the former genus Amoora). There are 250—300 species, occurring from China to the Pacific.

Aglais Urticae (small tortoiseshell butterfly; order Lepidoptera, family Nymphalidae). A butterfly in which the ground colour of the wings is a red-orange; markings are black, yellow, and blue. Adults overwinter and emerge in February or March. Larvae feed on nettles, feeding in groups until the last instar. There are several subspecies. The species is found throughout Europe, including Britain, and in temperate Asia.

Aglomerular. Lacking glomeruli, with reference to kidneys or kidney tubules.

Aglycone. Glycoside.

Aglyphic. Having solid, conical teeth, with reference to snakes.

Agnatha. A class of marine and freshwater vertebrates that lack jaws. They are fishlike animals with cartilaginous skeletens and well-developed sucking mouthparts with horny teeth. The only living agnathans are lampreys and hagfishes (order Cyclostomata), which are parasites or scavengers. Fossil agnathans, coverved in an armour of bony plates, are the oldest known fossil vertebrates. They have been dated from the Silurian and Devonian periods, 440—345 million years ago.

Aganathous

1. Lacking jaws.
2. Pertaining to the Agnatha, a class of primitive, jawless vertebrates which includes the ostracoderms and cyclostomes.

Agnosia. Loss of the ability to perceive or recognize persons or things.

Agonidae (poachers; class Osteichthyes, subclass Actinopterygii, order Scorpaeniformes). Family of small, marine, cold-water fish, with an elongated body, and narrow tapering tail. The pogge (Agonis cataphractus) of W. Europe reaches a maximum length of 20 cm. There are about 20 genera.

Agonis (family Myrtaceae, subfamily Leptospermoldeae)' Genus of shrubs and trees with alternate leaves, and five-petalled, usually white flowers. Agonis flexuosa (willow peppermint) is widely cultivated. There are 12 species, endemic to Australia, where they are found in areas of low fertility, including eucalputus forests, mainly in the west.

Agonists. Certation.

Agonism. Agonism.

Agonist. A primary muscle which, upon contracting, moves a part.

Agonistic. Describes behaviour between two rival individuals of the same species that may involve aggression, threat, appeasement, or avoidance, or that may be ritualized (*e.g.* as bird song, or as certain forms of display). Agonistic behaviour often arises from a canflict between aggression and fear.

Agonistic Behaviour (agonism). A type of animal behaviour which exhibits features of both aggression and avoidance.

Agonomycetales (Mycelia Sterilia). An order of the Hyphomycetes containing those mycelial imperfect fungi that do not produce spores, although some develop sclerotia. It contains about 200 species and 28 genera including Rhizoctonia, certain species of which cause damping-off disease, and Sclerotium.

[1]**Agouti**. A small South American rodent (Dasyproeta) of the family Dasyproctidae.

[2]**Agouti**. A pattern of coloration common in wild animals, especially rodents, in which individual hairs are mostly black with a narrow yellow band near the tip.

Agouti. A South American rodent looking like a long-legged guinea pig. There are several different kinds, all brownish in colour and about 50 cm long. Agoutis live mainly in woods and feed on leaves, roots and fallen fruit.

Agranulocyte. A white blood cell (leucocyte) that lacks granules in its cytoplasm. There are two types : lymphocytes and monocytes (comprising 25% and 4%, respectively, of all leucocytes).

Agrestal, Agrestis (latin). Growing in cultivated land, but not itself cultivated, *e.g.* a weed.

Agric Horizon. Mineral-soil diagnostic horizon formed from accumulation of clay, silt, and humuns, which has moved down from an overlying, cultivated soil layer. It is a soil horizon created by agricultural management, and is identified by its near-surface position, and by colloids accumulated in the pores of the soil. (USDA term.)

Agrimonia (family Rosaceae). Genus of perennial herbs with pinnate leaves having small leaflects alternating with the larger pairs, and long racemes of five-petalled flowers with many stamens, the receptacle bears hooked spines that persist and become woody in fruit, as does the deeply-hollowed receptacle. There are about 15 species, occuring in the nothern temperate zone, and in S. America.

Agrimony. Two species of flowering plant related to the rose and found mainly in dry grassy areas. Their yellow flowers are arranged to form a thin spike. The unrelated hemp agrimony belongs to the daisy family and is tall with clusters of pink flowers.

gr omyzidae (leaf-mining flies; class Insecta, order Diptera, suborder Cyclorrapha). Relatively large family of small, phytophagous flies, which are difficult to define. The adults are small, with the costa of the wing broken where Sc or R_1 runs out. The females have an oviscapt. The third antennal segment is short, rounded, and rarely has a sharp point. The arista is hairy or bare. The lower fronto-orbital bristles are incurved; vibrissae are present; and the tibia is without pre-apical bristles. The costal vein is incomplete, being broken some distance from its junction with the upper margin of vein 1, and continued as far as the apex of vein 3. The ovipositor has a non-retractile sheath comprising the fused seventh abdominal tergite and sternite. The larvae are leaf-miners, stem-borers, or gall-causers; and are sometimes pests of garden plants and some trees, although they are of no great economic importance. Those of Phytomyza thcts cause the familiar yellow blotching of holly leaves. They have a worldwide distribution, and 2300 species have been described.

Agriochoeridae (order Ruminantia, infra-order Tylopoda). Extinct family of N. American ruminants closely related to tne pig-like oreodonts, but more slightly built, and with more primitive features *e.g.* a long tail and clawed feet. They lived from the Upper Eocene into the Miocene, and, judging by their claws, inhabited trees or dug for roots and tubers.

Agrobacterium (family Rhizobiaceae). Genus of bacteria most of which are capable of causing the formation of galls in plants. No species can fix atmospheric nitrogen. A. tumefaciens causes crowth gall in many types of plant and is a serious plant pathogen in many parts of the world. These bacteria are found also in soil.

Agroforestry (agriforestry). System of mixed arable farming and forestry, recently designed for tropical developing areas, especially rainforest environments. The scheme is based on traditional shifting-cultivation methods; but in the fallow woodland phase, land is planted with a commercial crop, *e g.* rubber, instead of being abandoned to regenerate naturally. Alternatively, tree crops are interspersed with food croy patches on a continuous basis. The system ensures a continuous food supply, some continuous economic return, and the avoidance of soil degradation by exposure to tropical sunlight and rainfall during prolonged clearance; and it is suitable for small-scale family farmers.

Agrostology. The study of grasses.

Agynic

1. Lacking pistils.
2. Free from the ovary, said of pistils.

Ai. The three-toed sloth (Bradypus tridactylus).

Aianthous. Blooming continuously; having persistent flowers.

Aigsette

1. *ornith.* A plume or tuft, especially the beautiful plume borne by egrets during the breeding season.
2. *bot* The feathery crown of a seed, as that of a dandelion.

Aimophila (American sparrows; family Emberizidae). Genus of mostly nondescript, brown, ground-feeding seed eaters which nest on or close to the ground. There are 14 species, found in N. and C. America, with two species in S. America.

Air Bladder

1. A swollen air-filled region of the thallus seen in many algae of the order Fucales, notably Fucus vesiculosus (bladder wrack). The air bladders serve to increase buoyancy and may also play a part in respiration.

2. Any of the air-filled protuberances that develop by separation of the exine layers in the pollen grains of certain gymnosperms, *e.g.* Pinus. They give the pollen a characteristic winged appearance and assist the wind dispersal of the pollen.

Air Breathing in Fish. A number of tropical freshwater fish possess accessory breathing organs which enable them to acquire atmospheric oxygen and stay out of water for some time. In some fish, *e.g.* Anabas testudineus (climbing perch) and Clarias lazera (cartfish), the air taken up reaches richly vascularized chambers formed by modifications of the pharynx or the gill chambers. In some loaches even parts of the intestines are used for respiration. Other fish, such as lungfish, can breather air because of the presence of primitive lungs.

Air Cavity, Air Space

1. Used for a large intercellular space in a leaf into which a stoma is opening.
2. A cavity in the upper surface of the thallus of some liverworts which is opening externally by an airt-pore and containing chains of photosynthetic cells.
3. A large intercellular space in which air is sotred in some water-plants.

Air Chamber. Any of the air-filled cavities beneath the upper epidermis of the gametophyte thallus in many liverworts of the order Marchantiales. Such air chambers increase the surface area of the internal photosynthetic cells and so facilitate gaseous exchange. At the same time they maintain a humid atmosphere around these cells, thus reducing transpiration. Connection of the air chambers with the external atmosphere is facilitated by pores in the upper surface of the thallus.

Air Layering. A form of vegetative propagation often used to multiply greenhouse and indoor plants. A branch is stimulated to produce roots while still attached to the parent plant by making a shallow cut or removing a narrow ring of bark just below a bud. The cut is dusted with rooting powder containing synthetic auxins and kept open by inserting sphagnum moss. The whole area is then kept moist by wrapping the stem in moss and surrounding this with a plastic sleeve. Root formation is slow and may take up to two years.

Air Plant. Epiphyte.

Air Sac

1. In birds, one of a number of thin-walled sacs attached to the smaller bronchial tubes, some extending into the visceral spaces and some continuing as air cavities into the larger bones.

2. In insects, a thin-walled, distensible dilatation of a tracheal tube.

3. In mammals, an alveolar sac of the lungs.

Airway. *ornith.* A well-established route followed by migratory birds, especially waterfowl, in their seasonal migrations.

Ajacicyathida (class Monocyathea (Regulares). Order of Lower and Middle Cambrian archaeocyathids. They were usually solitary. The cylindrical of concial cup consists of two porous walls, the pores often being complex. The intervallum contains plane septa, or septa and tabulae, or radial links. Dissepiments or synapticulae are sometimes developed. The order contains four suborders and 5o families.

Ajuga (family Labiateae). Genus of herbs many of which have radical leaves forming a rosette. The leaves are simple or lobed, sometimes toothed, without stipules. The flowers are in at least two or many flowered whorls, often forming a terminal inflorescence. They are bisexual, with five fused sepals forming a toothed bell shape. The corolla has a short upper lip, and a larger, three-lobed lower lip with a ring of hairs in the corolla tube. The stamens are attached to the corolla tube. The ovary is superior, with two bilocular carpels aud four ovules. The fruit is an oval nutlet. There are approximately 45 species, found throughout the world in a wide variety of habitats.

Akaryote

1. Lacking a nucleus.
2. A non-nucleated cell.

Akaryote. Describing the stage in the nuclear cycle of the Plasmodiophoraceae, before meiosis in which little or no chromatin is visible in the nucleus. Sometimes applicable to the same stage in the life-cycle of any lower plant.

Akene. Achene.

Akinete. A sporelike resting stage in certain green algae in which the newly formed cell is fused to the wall of the parent cell.

Ala pl. **Alae**

1. A wing or winglike process.
2. A longitudinal fin or finlike structure present on certain nematodes.

Alangiaceae. Family of small trees, whose affinities are obscure The leaves are exstipulate, alternate, and simple. The flowers are cymose, regular, and bisexual. There are 4—40 sepals, petals, and stamens. Petals are sometimes basally fused. The ovary may be superior or inferior, and unilocular; with one pendulous ovule, and a disc present. The fruit is a drupe, and has a fleshy endosperm. The family is tropical. Alangium contains about 20 Old World species; Matteniusa contains three species found in northern S. America.

Alanine. A simple nonpolar amino acid in which the R group is CH_3 (illustration at amio acid). Alanine is formed in a transaminase reaction in which glutamine donates an amino group to pyruvic acid to form alanine and α-ketoglutaric acid. This reaction is reversible; alanine can be deaminated to pyruvate and subsequently oxidized in the TCA cycle. Along with aspartic acid, serine, glutamic acid and glycine, alanine is one of the early products of photosynthesis.

Alar. Pertaining to, resembling, or shaped like a wing.

Alarm Response. Signals emitted by an animal that serve to warn others of a danger the individual has perceived. The signals may be visual (*e.g.* the white tail of a rabbit, displayed when running), aural (*e.g.* the cell of a number of a flock of ground-feeding birds), or olfactory (among some fish and invertebrates).

Alarm Substance. Substance released from the skin of injured fish. It causes an immediate alarm reaction among fish nearby, leading to a breaking up of a school formed by members of the same species. Originally discovered in minnows and other Cyprinidae, the alarm substance is probably present in several fish families.

Alate

1. Bearing wings or alae.
2. *bot.* Bearing winglike processes.

Alauidae (larks; order Passeriformes). Family of small, grey-brown and buff birds, their clours adapted to suit their environments. They have long, sharp, hind claws that help them to walk and run rather than hop. They are renowned for their song, delivered from a perch or in flight. They are seed and insect eaters. There are 15 genera, with 77 species, found in most of the Old World, with Eremophila alpestris (shore lark or horned lark) found also in N. America and Mexico.

Albacore. The tuna or tunny, q.v.

Albatross. A group of birds related to the petrel, and characterized by tubelike nostrils on top of the beak. Albatrosses are large birds, with goosesized bodies and very long, narrow wings. The wandering albatross has a wing span of 3.5 metres or more—greater than that of any other bird. Most live in the southern hemisphere and they spend nearly all their lives at sea, gliding gracefully over the waves.

Albedo. The ratio of the intensity of light reflected from a surface to the intensity of light received. All leaves reflect or transmit infrared radiation but different species reflect different amounts of visible radiation. Thus xerophytic plants tend to reflect more light than mesophytic plants.

Albinism. Lack of pigment in the skin, resulting in pure white individuals. Albino mammals and other vertebrates have pink eyes because there is no pigment in the iris to mask the colour of the blood capillaries.

Albino. An animal that lacks skin pigments, resulting in a pure white individual. The pink eyes associated with albinism are due to the lack of pigment in the iris which allows the colour of the blood show through.

Albizia (family Leguminosae, subfamily Mimosoideae). Genus of tropical plants, sometimes known (incorrectly) as Albizzia. The subfamily Mimosoideae was ranked formerly as a family, Mimosaceae. It comprises shrubs and trees with feathery leaves and fluffy yellow flower heads, similar to Acacia (to which the genus is closely related). Several species are valuable timber trees. There are 10-12 species, occurring in Australia.

Albuginaceae (order Peronosporales). Family of fungi which are obligate parasites of flowering plants, causing diseases known as 'white rust' or 'white blister'. Sporangiophores develop beneath the host epidermis, causing it to swell and burst. The white powdery sporangia are then dispersed by wind. There is one genus.

Albumen

1. Egg white, consisting principally of albumin.
2. Stored food in the seed of a plant.
3. Albumin.

Albumin (albumen). One of a group of globular proteins that are soluble in water but form insoluble coagulates when heated. Albumins occur in egg white, blood, milk, and plants. Serum albumins, which constitute about 55% of blood plasma protein, help regulate the osmotic pressure and hence plasma volume. They also bind and transport fatty acids, α-lactalbumin is one of the proteins in milk.

Albuminoids A general term used for proteins.

Albuminous. Containing endosperm at maturity, with reference to seeds.

Albuminous Cell

1. A specialized parenchyma cell, physiologically and anatomically associated with a sieve cell in gymnosperm phloem. Albuminous cells are analogous to the companion cells of angiosperm phloem but unlike companion cells they do not usually arise from the same mother cell as the sieve element.

2. An albumen-containing cell, found in certain seeds.

Albuminous Seed. Endosperm.

Alburnum. An obsolete term for sap-wood.

Alcedinidae (kingfishers; order Coraciiformes). Family of brightly coloured birds with large heads, short necks, and compact bodies. They have short, rounded wings, and a short tail. The bill is long, straight, and massive; the toes syndactyl. The sexes are usually alike. They are found in riverine and terrestrial habitats, feeding on fish, insects, and small vertebrates. There are 14 genera, with 84 species, of which the best known are Alcedo atthis (common kingfisher) and Dacelo novaeguineae (kookaburra). They are found worldwide.

Alcedo (kingfishers; family Alcedinidae). Genus of small kingfishers, with blue upper-parts, and chestnut or white under-parts. A brilliant blue line down the back and rump shows in flight. They are fish-eating ond riverine, nesting in burrows. There are nine species, found throughout African, Palaearctic, and Oriental regions, and also in New Guinea.

Alchimilla (lady's mantles; family Rosaceae). Genus of herbs and low shrubs in which the stipules are leaf-like; the leaves alternate, palmately-veined or lobed; and the flowers green, small, with four sepals, borne in cymes, with no petals. An epicalyx of four lobes is present. There are four or five stamens. The fruit is dry, and formed from the single carpel. Many apomictic species occur in Europe, and several local endemics in E. Africa. They occur in open, grassy habitats. Some are cultivated. There are about 200 species, found in the northern temperate zone, and in the mountains of tropical Africa.

Alcian Blue. A stain that enhances the preservation of cell surface materials, especially mucopolysaccharides. It is usually used during fixation with glutaraldehyde for transmission electron microscopy.

Alcidae (auks; order Charadriiformes). Family of mainly black and white, small-winged, diving seabirds. The legs are set well back; the feet are webbed. Bills vary from long and pointed to laterally compressed and high. Auks are mainly pelagic and gregarious, breeding in burrows or crevices, or on open cliff ledges, usually colonially. Brachyramphus marmoratus (marbled murrelet) breeds on forest branches. Auks feed on fish and invertebrates. Pinguinus impennis (great auk) is extinot. There are 12-14 genera, with 22 species, found in northern regions of the Pacific and Atlantic, and in the arctic.

Alcippe (fulvettas, nun babblers; family Timaliidae). Genus of small, short-billed birds, brown and grey in colour, some marked with chestnut, yellow, black, and white. They are generally salitary, and most inhabit forests. They are insectivorous, but also feed on snails and berries. There are 17-18 species, found mainly in SE Asia, with two species occurring in Africa.

Alcohol. An organic compound containing a functional hydroxyl (OH) group. Grain alcohol or ethanol (C_2H_5OH) is a colorless, volatile, inflammable liquid produced in alcoholic fermentation. It is also obtained from distillation of fermented grain or saccharine liquids or produced synthetically. In biology it is widely used as a solvent, preservative, fixative, and dehydrating agent.

Alcoholic Fermentation. A form of anaerobic respiration in which glucose is broken down to form ethanol and carbon dioxide. It is carried out by yeasts and various other fungi and by certain bacteria. Fermentation takes place outside the organism and is catalysed by enzymes of the zymase complex. These are either secreted by living cells or released on cell death. Fermentation usually stops due to cell poisoning when the alcohol level reaches about 15%. The process is central to the brewing, wine-making, and baking industries. Since free oxygen is not available as a hydrogen acceptor acetaldehyde is used instead. Pyruvic acid, formed by glycolysis, is broken down to acetaldehyde and carbon dioxide. The acetaldehyde is then reduced by $NADH_2$ to form ethanol and NAD. The process yields about 72 kcals from each glucose molecule. This is only about 10% of the energy that would be released by complete oxidation of glucose, as in aerobic respiration.

Alcyonacea (soft corals; subclass Octocorallia). Order of corals which appeared first in the Lower Jurassic and which contains the modern genus Alcyonium (dead man's fingers). Members of the order are characterized by having retractable polyps with eight branching tentacles; the polyps are embedded in the body mass, which has a skeleton of numerous, free, calcareous spicules. Usually these corals are attached to rocks. The order contains six families, with 36 genera.

Alcyonarian. Any anthozoan of the subclass Alcyonaria which includes the precious corals, sea fans, and sea feathers.

Alder. Member of a group of deciduous trees, belonging to the birch family, that thrive in wet places. The trees grow

Alder trees

about 25 metres tall. The flowers are arranged as catkins and open before the leaves. The female catkins are woody and barrel shaped. The common alder has dark, almost black bark and rounded leaves. The grey alder has grey bark and toothed triangular leaves.

Alderfly. An insect of the order Megaloptera (sometimes considered as a suborder of Neuroptera) with carnivorous, aquatic larvae. Most species belong to the genus Sialis.

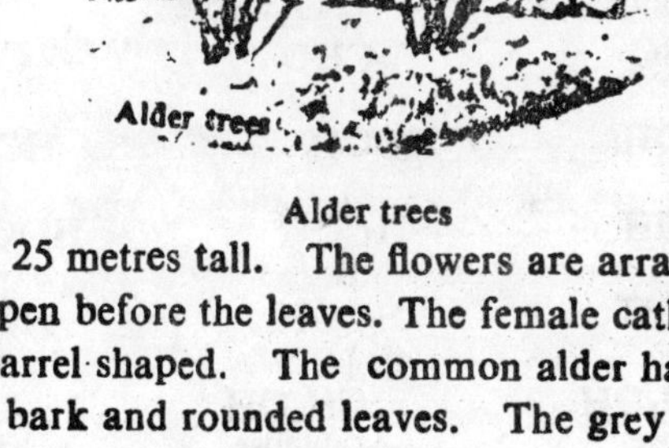

Alder fly

Aldohexose. An aldose sugar having six carbon atoms.

Aldopentose. An aldose sugar having five carbon atoms.

Aldose. Any monosaccharide with the carbonyl (CO) group on the terminal carbon, so forming an aldehyde (CHO) group. The simplest aldose is the three-carbon sugar glyceraldehyde. Other aldoses include glucose, galactose, and mannose, which are aldohexoses, and ribose, arabinose, and xylose, which are aldopentoses (diagram). Compare ketose.

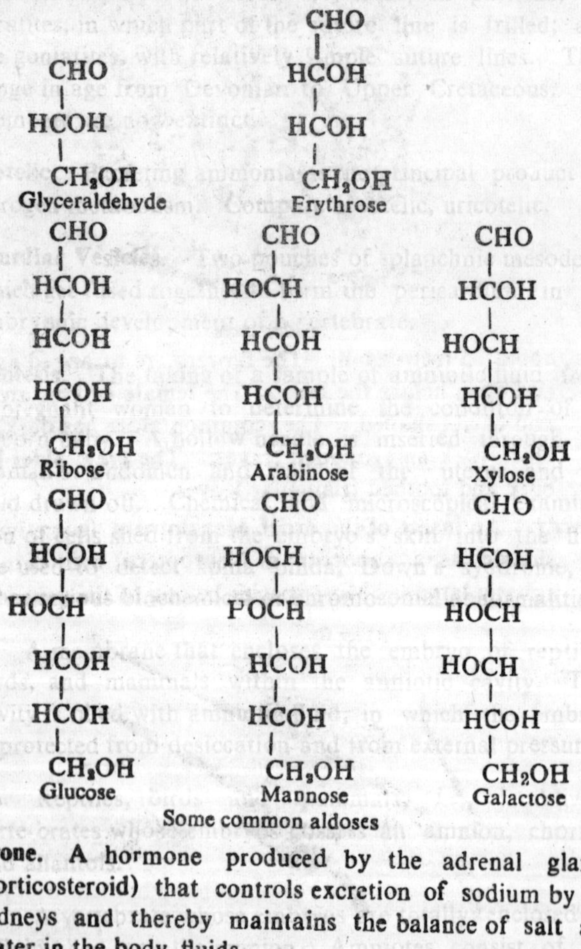

Some common aldoses

Aldosterone. A hormone produced by the adrenal glands (corticosteroid) that controls excretion of sodium by the kidneys and thereby maintains the balance of salt and water in the body fluids.

Alecithal. Possessing little or no yolk, said of ova as those of placental mammals.

Alectoris (partridges; family Phasianidae). Genus of medium-sized, rotund, short-tailed game birds with red bills and legs, and strikingly barred flanks. They prefer to run rather than fly, are ground-nesting, and inhabit mountainous regions and open country. There are seven species, found in Europe, Asia, and N. Africa. A. rufa (red legged partridge) has been introduced in Britain.

Aleppo Gall. A tumourous growth taking place on various Oak species. It has gallic acid, and gallitannic acid. Extract finds use in tanning leather.

Aleuria (order Pezizales). Genus of fungi which typically form large, saucer-like, bright orange or yellow apothecia on the ground in woods, fields, etc. A. aurantia is the commonest of the three British species, and is found on soil in woods, lawns, grass verges, etc. in autumn.

Aleuriospore. A terminal, thick-walled spore. It is formed from the end of a hypha, from which it gets separated by a cross-wall.

Aleurites (family Euphorbiaceae). Genus of trees producing fruits that are large and round, with a thick, fleshy pericarp, and that contain a few big, very hard, oily seeds. A. moluccana (candlenut), of Malesia provides oil for candles, soaps, paints, and varnishes. A. fordii and two relatives (sometimes called Vernicia), of China, provide tung oil for varnishes and paints. There are five species, occurring from tropical Asia to the Pacific.

Aleurone. Protein occurring as small particles in the endosperm of seeds of various plants, especially cereals.

Aleurone Grains. Granules of protein occur in cells. These cells often form a distinct layer called the aleurone layer, *e.g.*, in the endosperm of Maize.

Aleurone Layer (proteinaceous endosperm layer). The proteinaceous outermost layer of the endosperm in seeds of the Gramineae (grasses) and Polygonaceae. The cells of the aleurone layer contain aleurone grains, which store protein for later use by the embryo.

Aleuroplast. A type of proteoplast in which protein in the form of aleurone grains is stored. They are common in seeds; for instance, in the endosperm of castor oil.

Alexanders. A tall early-flowering plant of the carrot family which produces clusters of small yellow flowers and has glossy leaves. It is found mainly in hedgerows and on wastelands near the coast.

Alewife. A clupeid fish (Pomolobus pseudoharengus), an anadromous fish of the Atlantic coast and also common in the Great Lakes.

Alfalfa. A hardy crop plant, also known as lucerne, which is widely grown as animal feed. Like other members of the pea family, it restores nitrogen to the soil by way of its root nodules. Each leaf has three leaflets on a short stalk.

Alga. A group of simple plants, most of which live in water, others in damp places on land. About 25,000 different kinds are known, some green, some brown and some red. Many are single-celled, and too small to be seen without a microscope. Others are large seaweeds. Algae can make food from sunlight by the process called photosynthesis, and in turn provide food for animals.

Algae. An extremely diverse group consisting predominantly of aquatic plants showing relatively little differentiation of tissues and organs as compared to bryophytes and tracheophytes. The group includes both prokaryotic and eukaryotic organisms ranging from unicells through colonial and filamentous forms to parenchymatous seaweeds over 50 m long. Some 23 500 species of eukaryotic algae and some 1450 species of prokaryotic algae have been described. Adaptation to an aquatic environment has led to the development of many distinctive biochemical traits. Algal sex organs are usually unicellular.

Formerly the algae were placed with the bacteria and fungi in the division Thallophyta. Increasing appreciation of the very basic differences between such organisms and between the algal groups themselves has led most systematists to consider each group as a separate class or division. The groups are classified according to their pigments, food reserves, cell-wall materials, number and types of flagella, and ultrastructural details.

Algal Bloom. Sudden growth of algae in an aquatic ecosystem. It can occur naturally in spring or early summer when primary production exceeds consumption by aquatic herbivores. Algal blooms may also be induced by nutrient enrichment of waters due to pollution. They are a characteristic symptom of eutrophication.

Algin. Sodium alginate, a purified carbohydrate obtained from brown seaweeds, widely used for its water-binding, thickening, and emulsifying properties.

Alginic Acid. A carbohydrate polymer consisting of D-mannuronic acid and L-glucuronic acid units. It is found principally in the cell walls of brown algae where it functions as an ion exchange agent. Alginic acid derivatives have commercial value; they are used in the making of fireproof and disposable fabrics, as emulsifying agents, in some types of surgical suture, and as stabilizers in food.

Alien. A plant which has been introduced by man, and then naturalized.

Aliferous. Possessing or bearing wings.

Aliform. Resembling a wing.

Alima. The larval stage of stomatopod crustaceans.

Alimentary

1. Pertaining to food or nutrition.
2. Nourishing or nutritious.

Alimentary Canal (digestive tract; gut). A tubular organ in animals that is divided into a series of zones specialized for the ingestion, digestion, and absorption of food and for the elimination of indigestible material.

The wall of the alimentary canal of vertebrates is made up of several of coats. Starting from the inside they are :

1. *Mucous membrane*—epithelium and underlying connective tissue.
2. *Muscularis mucosa*—a thin band of muscle.
3. *Submucous coat of areolar connective tissue with blood vessels and lymphatics.*
4. *The muscle coat*—circular and longitudinal, and in the stomach, oblique.
5. *The peritoneum*—a thin layer of connective tissue with an outer squamous epithelium.

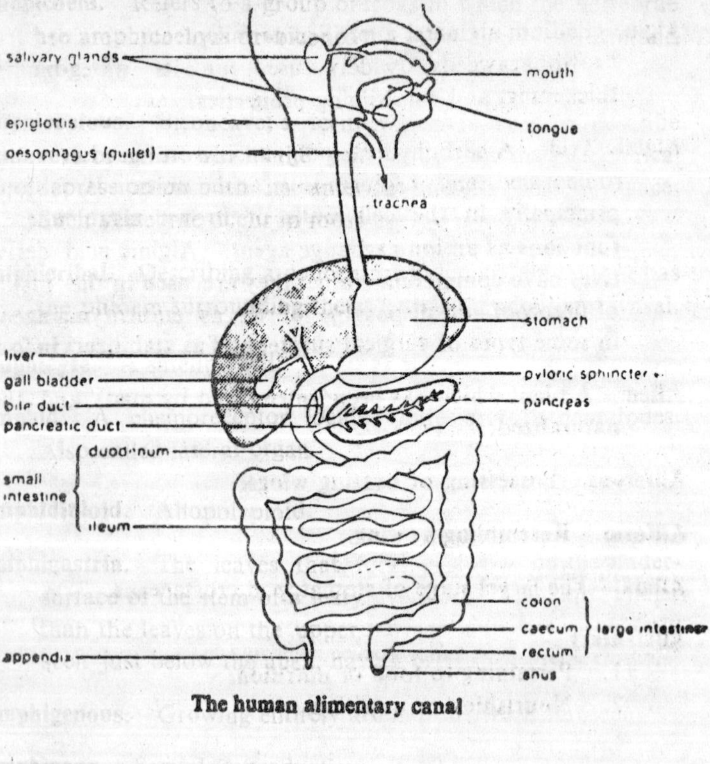

The human alimentary canal

Alkali. A substance which neutralizes an acid, usually a metallic hydroxide; a base.

Alkaline Soil. A soil having alkaline pH. It is formed in areas of very high rainfall or dryness due to depositon of calcium sals in surface layers or on the soil.

Alkaloids. A class of nitrogen-containing usually basic plant products, which are often poisonous. Many alkaloids *e.g.* morphine codeine, nicotine, and cocaine, have been utilized in medicine and other fields. Their natural functions are not well understood. Some, through their bitterness and toxicity, may serve to protect the plants from herbivorous animals. Others are though to be in volved in nitrogen metabolism (*e.g.*, nicotine is involved in the absorption of nitrate through plant roots). Many are thought to be simply end products of nitrogen metabolism, which are stored in leaves fruits, and flowers before being discarded. Certain families, *e.g.*, the Solanaceae, Leguminosae, and Papaveracease contain many alkaloidal species while monocotyledons rarely produce alkaloids. Most groups of lower plants contain alkaloidal species though alkaloids have not been found in the algae. Three classes of alkaloids have been recognized.

Alkalosis. A condition in which there is an increase in alkaline substances in the blood, especially sodium bicarbonate.

Alkanet. Several specics of mainly blue-flowered plant with hairy leaves and stems, belonging to the borage family. The green alkanet has leafy clusters of blue flowers on long stalks. It is often found in woods and hedgerows.

Alkaptonuria. A metabolic condition characterized by the presence of alkapton (homogentisic acid) in the urine; inherited as a mendelian recessive trait.

Allantoic Acid. An excretory product of many vertebrate embryos, which is formed by the oxidation of allantoin.

Allantoin. An excretory product of many vertebrate emoryos which is formed from breakdown of uric acid.

Allantois. Used for describing a sac or bladder which grows out from the hind-gut in the embryos of reptiles birds and mammals. As it enlarges it gets very vascular, extending outside the embryo, and is the chief organ by which dissolved food and oxygen get absorbed.

Allantospermum (family Simaroubaceae). Genus of biggish trees in which the leaves are simple; the flowers are bisexual, and pentamerous; and the fruit is a capsule. There are four or five species occurring in Boreo, Malaya, and Madagascar : a remarkable disjunction in range.

Allantoic Bladder. In vertebrates, a sac formed from the posterior region of the alimentary canal, functioning as a urinary bladder in amphibians, receiving metabolic wastes in embryonic reptiles and birds, and forming part of the placenta in Eutheria.

Allantois. One of the membranes that develops in embryonic reptiles, birds, and mammals as a growth from the hind-gut. It acts as aurinary bladder for the storage of waste excretory products in the egg (in reptiles and birds) and as a means of providing the embryo with oxygen (in reptiles, birds, and mammals) and food (in mammals placenta).

Allele (allelomorph). A form in which a gene may occur. Different alleles of a gene give rise to different expressions of a character. Hence alleles for 'green' and 'yellow' are alternative expressions of the gene governing the characteristic for seed colour. Alleles of a gene always occupy the same site (locus) on homologous chromosomes. A diploid organism whose cells contain two identical alleles is said to be homozygous. One with two different alleles at a locus is said to be heterozygous. The allele expressed in a heterozygous organism depends on the dominance relationship between the two alleles. If one is expressed to the exclusion of the other it is described as dominant and the latter recessive. If the heterozygote is intermediate in appearance between the two homozygotes, the alleles are said to exhibit incomplete dominance.

Allelism. The relationship between alleles.

Allelomorph. An allele.

Allelopathy. The release of a chemical by a plant that inhibits the growth of nearby plants and thus reduces competition. For example, pines (Pinus) produce substances that kill any seedlings of the same species (autotoxicity or autoallelopathy) growing too close to the parent plant. Allelopathy may also be indirect, if one plant inhibits the growth of a second plant or microorganism that itself is essential to the growth of a third plant. Allelopathic substances have been shown to be responsible for various changes during plant successions. For example, in a field succession the pioneer weed stage is replaced by annual grasses because the weeds produce substances inhibitory to other weeds.

In arid or semiarid habitats the allelopathic substances are often volatile terpenes released from the leaves. For example, in the Californian chaparral aromatic shrubs, such as Artemisia and Salvia, release oils (*e.g.* camphor) that inhibit the germination of herbaceous plants. In wetter areas various phenolic compounds have been shown to be allelopathic.

Allen's Rule. The disputed idea that a race of warm-blooded species in a cold climate typically has shorter protruding body parts (nose, ears, tail, and legs) relative to body size than another race of the same species in a warm climate. The varying wing lengths in gulls are said to support this view. The hypothesis is that long protruding parts emit more body heat, and so are disadvantageous in a cool environment, but advantageous in a warm environment. Critics point to many other adaptations for heat conservation which probably are more important, notably fat layers, feathers, fur, and behavioural adaptations to avoid extreme temperatures.

Allergen. An agent capable of inducing a state of allergy, as pollen, drugs, dust, animal hair, and certain foods, especially proteins.

Allergy. An antigen-antibody reaction occurring in certain individuals upon exposure to substances which under similar conditions are innocuous to other individuals. The exciting agents, called allergens, act as antigens causing the production of specific antibodies in sensitive persons. Common manifestations are hay fever, asthma, urticaria, gastrointestinal disturbances, and serum sickness.

Alligator. A large reptile closely related to the crocodiles, but distinguished by its blunter shout and by the fact that the fourth tooth in the lower jaw is not visible when the mouth is shut. Alligators live in warm rivers and spend much of their time basking on the banks. Adults feed on fish, birds, and small mammals that come to the river to drink or swim. There are two species of alligator, one in North America reaching 4.5 metres or more in length, and a much smaller one in China.

Alligator Gar. The gar.

Allium (onions, garlic; family Liliaceae). Genus of bulbous herbs, all of which contain oils with a pungent onion or garlic odour. The leaves are undivided, parallel-veined, and all radical. The flowers are borne in umbels enclosed at first in a spathe, which later splits. The perianth consists of six free, equal segments, and six stamens. Bulbils often partly replace the flowers. The fruit is a three-called capsule. Many species are cultivated as vegetables, or for their flowers. There are more than 500 species found in the north temperate zone, south to Mexico and Ethiopia.

Allocarpy. When fruiting takes place after cross-fertilization.

Allochronic. Not occurring at the same time level; not contemporary, said of species.

Allochronic Speciation. The formation of new species during the passage of time, tending to form a gradation from one species to the next.

Allochthonous

1. Brought in from an outside source, as sediment in a lake.
2. Exotic; imported or having migrated from another area, said of species.

Allocryptic. Concealing; designating animals which cover themselves for concealment.

Allogamy (cross fertilization, exogamy). Fusion of female and male gametes derived from genetically dissimilar individuals of the same species. It promotes the recombination of variable genetic material so that the total population shows greater variation than an autogamic population and has greater adaptive potential. Various methods have developed in the plant kingdom to encourage or necessitate allogamy (dichogamy, dioecious, monoecious, incompatibility, chasmogamy, heterostyly). When cross fertilization is obligate, useful genetic traits may be diluted and isolated individuals cannot reproduce.

Allograph. A homograph.

Allometric Growth. The regular and systematic pattern of growth such that the mass or size of any organ or part of a body can be expressed in relation to the total mass or size of the entire organism according to the allometric equation: $Y=bx^{\alpha}$, where Y=mass of the organ, x=mass of the organism, α=growth coefficient of the organ, and b=*a* constant.

Allometry

1. The phenomena of relative growth; heterauxesis.
2. A study of changes in shape or proportions with increase in size.

Allopatric. Describing a population or species that is unable to breed with a related group of organisms because of geographical separation by distance or by natural barriers, such as water, mountain ranges, deserts, etc.

If two groups that have been separated in such a way develop different adaptations such that they would not be able to interbreed even if the barriers broke down then allopatric speciation is said to have occurred. Compare sympatric.

Allopatric Speciation. Formation of new species from the ancestral species as a result of the geographical separation or fragmentation of the breeding population. Separation may be due to climate change, causing the gradual fragmentation of the population in a few surviving favourable areas (*e.g.* during glaciation or developing aridity), or may arise from the chance migration of individuals across a major dispersal barrier. Genetic divergence in the newly isolated daughter populations ultimately leads to new species; divergence may be gradual or, according to punctuationist models, very rapid.

Allopatry. Describing a population or species where two groups of isolated organisms can not breed with each other due to their separation by distance or natural barriers *e.g.*, water, deserts or mountain range. Population distributed in such a way (allopatric) may develop different adaptations of different habitats and then they may fail to interbreed even if barriers are broken (allopatric speciation).

Allopelagic. Pertaining to aquatic organisms found at various depths.

Allophilic. Used for flowers of an insect pollinated species that can be pollinated by different types of insects.

Allopolyploid. A polyploid arising from hybridization of two species or genera with resulting duplication of chromosome complement.

A lopolyploidy. A form of polyploidy that results from the combination of sets of chromosomes from two or more different species. A (diploid) interspecific hybrid is normally sterile because there is only one of each kind of chromosome per cell. Thus at meiosis no bivalents are formed and any resulting gametes are usually inviable as they contain either too many or too few chromosomes. However if some unreduced diploid gametes form and fuse then the resulting tetraploid will usually be fertile as its nuclei will contain pairs of homologous chromosomes. An allotetraploid may cross with a diploid to form a sterile triploid hybird, which, if it produces unreduced triploid gametes, may give rise to a fertile allohexaploid. If an allohexaploid crosses with a diploid an allooctaploid could arise in a similar fashion, and so on to higher levels of allopolyploidy. An allopolyploid is often intermediate in appearance between both parental species and cannot reproduce with either. Hence it may merit the status of a new species. Allopolyploidy has been described as instant evolution. Many crop plants are believed to have originated in this way. Wheat (Triticum aestivum), for example, is an allohexaploid. It has a chromosome number of 42 and is probably derived from the three species T. monococcum, Aegilops speltoides, and A. squarrosa, each with 14 chromosomes.

If at meiosis an allopolyploid forms only bivalents, *i.e.* it acts like a diploid, then it is termed an amphidiploid. Raphanobrassica (allotetraploid) is an example. However if certain of the chromosomes from the two parent species are sufficiently similar (homoeology) then multivalents may be seen at meiosis. When this occurs the allopolyploids are termed segmental allopolyploids. An example is the allotetraploid Primula kewensis derived from a cross between P. floribunda and P. verticillata. Sometimes segmental allopolyploids show reduced fertility as compared to amphidiploids and it is clear that in some allopolyploids there are mechanisms that prevent the formation of multivalents. Wheat, for example, normally behaves as an amphidiploid but if chromosome V of the B genome (the genome derived from A. speltoides) is missing then multivalents are formed.

All-or-none Law. A principle in physiology to the effect that an irritable structure, when stimulated, responds to its fullest extent or not at all. It applies to skeletal muscle (a single fiber), cardiac muscle (the heart as a whole), and a single nerve fiber.

All-or-none Response. A type of response that may be either complete and of full intensity or totally absent, depending on the strength of the stimulus; there is no partial response. For example, a nerve cell is either stimulated to transmit a complete nervous impulse or else it remains in its resting state; a stinging thread cell of a coelenterate is either completely discharged or it is not.

Allosematic. Having protective marks or coloration resembling that of dangerous or distasteful forms.

Allosome. Sex chromosome. Any chromosome other than a typical one.

Allosomal Inheritance. Inheritance of characters which are carried on an allosome.

Allosteric Effect. The binding of a ligand to one site on a protein molecule such that the properties of another site on the same protein are affected. Some enzymes are allosteric proteins, and their activity is regulated through the binding to an allosteric site of an effector.

Allosteric Enzyme. A regulatory enzyme responsive to alterations in the metabolic state of a cell or tissue. Its catatytic activity is modified by the noncovalent binding of a specific metabolite at a site (allosteric site) other than the active site. The most common type of allosteric enzyme is one found at the beginning of a multienzyme sequence that is inhibited specifically by the end product of the reaction sequence (feedback inhibition). However, the activity of an allosteric enzyme is not always decreased by the action of a modulator; some allosteric modulators increase the enzyme's activity. Allosteric enzymes may also have more than one modulator.

Reactions involving allosteric enzymes are always irreversible in the cell, as it would be impossible to regulate a reversible reaction. Allosteric enzymes are often more complex structurally than other enzymes; all known allosteric enzymes have at least two protein subunits.

Allosteric Site. A regulatory site on an allosteric enzyme where a specific effector or modulator can reversibly bind. When bound, the modulator either activates (a positive modulator) or inhibits (a negative modulator) the enzyme by changing its shape. Some allosteric enzymes have a site for only one modulator (monovalent enzymes) while others have several allsoteric sites (polyvalent enzymes).

Allosyndesis. Pairing of chromosomes derived from different species. It is seen in some diploid hybrids and in segmental allopolyploids (allopolyploidy).

Allotetraploid. An allopolyploid that has originated from an interspecific hybrid through the formation and fusion of unreduced diploid gametes. An example is Raphanobrassica, which is obtained by crossing Raphanus sativus (radish) with Brassica oleracea (cabbage). The F_1 is sterile but diploid gametes are sometimes formed, which on selfing give a fertile tetraploid F_2.

Allotropous

1. Having nectar readily available to insects, said of flowers.
2. Not restricted to certain kinds of flowers, said of insects.

Allotropous Flower. A flower which has the nectar freelyexposed and available to a wide variety of insects.

Allotype. A paratype of the opposite sex.

Allspice. The dried berries of the West Indian pimento, it tastes like a blend of cinnamon, clove and nutmeg.

Alluvial Soil. A type of azonal soil formed on the flood plains of river valleys and at river mouths (alluvial fans or deltas). New material is successively deposited on the surface when the land is subjected to flooding. In tropical regions such areas may be used as paddy fields, and in arid regions the extent of the alluvial soils may be artificially increased by irrigation methods, as along the banks of the Nile. The polders of the Netherlands are reclaimed marine alluviums.

Almond. A deciduous tree of southwestern Asia which has been grown in Europe for many hundreds of years. It belongs to the rose family. There are several varieties of

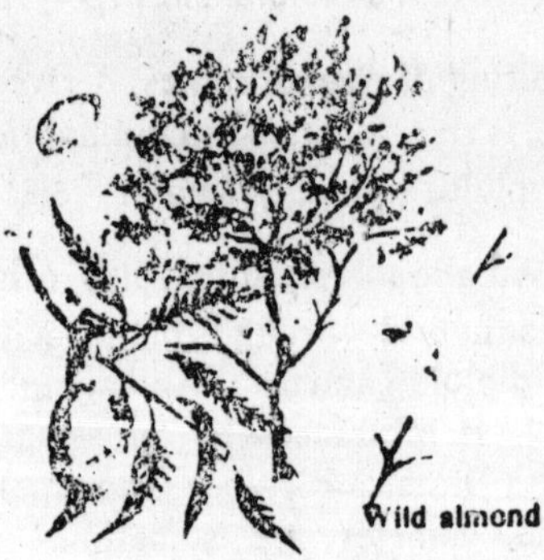

Wild almond

almond which can be divided into two groups—those producing sweet, edible nuts and those with bitter nuts. The trees are grown in warmer southern parts for their nuts but in NW Europe they are grown for their pretty pink blossom.

Alnus (alder; family Betulaceae). Genus of deciduous trees which are common it wet places. The roots have nitrogen-fixing nodules. The leaves are simple; the flowers are unisexual and brone in catkins, the female flowers developing into a woody, persistent, one-like structure. The trees are useful for timber and as ornamentals. There are 35 species, occurring in the north temperate zone, eptending south to the Himalayas and Andes.

Aloe. A group of perennial plants of southern Africa with stiff fleshy leaves and red or yellow flowers. Aloes range from a few centimetres to 30 metres tall. The leaves yield a bitter juice which is used foı medicinal purposes and fibre which is used for rope and cloth.

Alopecia A pathological disorder which is resulted by deficiency of panthothenic acid; characterised by falling of hairs.

Alpaca. A domesticated from of the guanaco with a long soft coat, normally either totally black or white. Alpacas are found in Bolivia and Peru. Their wool is clipped and made into expensive cloth.

Alpha Amino Acid. Amino acid in which the amino group is attached to the number-two, or 'alpha' carbon, adjacent to the carboxyl group. Compounds of this type represent the basic building blocks of peptides and proteins.

Alpha-cells (α-cells). A group of cells present in the islets of Langer-hans (constitute 20% of these structures) which have granules that stain red with Mallory stain. Secrete glucagon, a glycogenolytic hormone which converts glycogen to glucose.

Alpha Helix. The right-handed, or less commonly left-handed, coil like configuration of a polypeptide chain representing the secondary structure of some protein molecules, particularly the globular variety. The configuration is maintained through intrachain hydrogen bonding between >CO and >NH groups of peptide bonds.

Alpha-ketoglutarate Oxidase. An enzyme in the Citric Acid Cycle (Kreb's Cycle), which oxidises α-ketoglutarate to succinic acid, by oxidative decarboxylation.

Alpha keto Glutaric Acid (a-Keto Glutaric Acid). An organic compound which is formed during Kerb's cycle by dehydrogenation of isocitric acid and is changed into succinic acid by its own dehydrogenation. (Also see Kerb's cycle). Also formed in tissue by glutamic acid during transamination. (Glutamic acid+Pyruvic acid=α-Keto glutaric acid+Alanine).

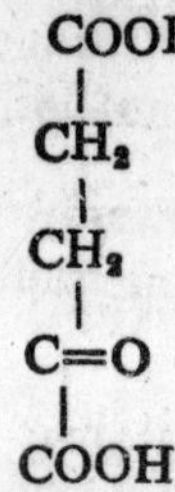

Alpha-Keto Glutaric Acid

Alpha-motoneurons. Refer to the neurons which are exiting the spinal cord and synapsing with extrafusal fibres. These initiate the muscle contraction.

Alpha-naphthol Test. A biochemical test to detect the presence of carbohydrates in solution, also known as Molisch's test (after the Austrian chemist H. Molisch (1856—1937), who devised it). A small amount of alcoholic alpha-naphthol is mixed with the test solution and concentrated sulphuric acid is poured slowly down the side of the test tube. A positive reaction is indicated by the formation of a violet ring at the junction of the two liquids.

Alpine. A major regional community (biome) of vegetation in high mountainous regions and on high-level plateaus. The plants generally grow in thin stony soil and are subjected to high light intensity and high wind speeds. The climate differs from that of the tundra as there is daylight in winter and darkness in summer, more precipitation and wind, and a higher degree of solar radiation, but no permafrost. The number of plant species, although limited, is greater than that found in the tundra. Certain drought-tolerant species (*e.g.* grasses, sedges, mosses, and lichens) are common to both communities. In temperate latitudes there are many brightly coloured flowers in summer on the alpine pastures.

The vegetation changes as the altitude increases due to the associated drop in temperature. The vegetation also differs between north- and south-facing slopes. The lower limit of the alpine zone (tree line) varies in different mountain regions and also varies with the wetness of the locality and the mass of the mountain range.

ALS. Antilymphocyte serum.

Alstonia (family Apocynaceae). Genus of small to big trees in which the crown is often pagoda-like, with distinct whorls of branches. The bole often has huge buttresses. The leaves are often whorled; the flowers are cymose. The fruits are paired, woody follicles, containing many seeds, and having a tuft of silky hairs at both ends. The timber is easily worked and is sometimes very light. The bark yields latex which is used in traditional medicine. There are 39 species, occurring in the Old World subtropics and tropics.

Alternanthera (family Amaranthaceae, tribe Gomphreneae). Genus of herbs and shrubs with a single ovule and bilocular stamens which distinguish them from other members of the large family. Some species are grown for their ornamental leaves. A sessilis has fleshy leaves which are eaten in some tropical countries. The genus is cosmopolitan but mainly subtropical, and six or seven species are endemic to Australia, where they are found in arid habitats.

Alternate. Describing a form of leaf arrangement in which there is one leaf at each node (illustration at phyllotaxis). This pattern is found in most plants. Compare opposite, whorled.

Alternate Host. A plant, other than the main host, on which a pathogen or pest can live. Alternate hosts, which are often weeds, can provide a means for the pathogen to survive when its main host is not available. For example, downy mildew of beet (Peronospora farinosa) can overwinter in wild beet. Hawthorn (Crataegus) is an alternate host and important inoculum source for the fireblight bacterium (Erwinia amylovora), which infects pears and apples. Many rusts also overwinter on alternate hosts. Certain insect pests overwinter as eggs on alternate hosts, *e.g.* blackfly (Aphis fabae) migrates to the spindle tree (Euonymus europaeus) in the autumn. Virus diseases may be found in weed hosts near crops. The term alternative host may be used when the pathogen has a number of different hosts. Control of the alternate hosts can be an important way of reducing inoculum sources of some diseases.

Alternation of Generations. The regular occurrence of two distinct forms during the life cycle of an animal or plant. The best examples are seen among tha ferns and their relatives. The mature fern scatters spores, but instead of growing into new fern plants the spores produce totally different plants known as prothalli. These are tiny green plates that grow unnoticed on the ground. Male and female cells are produced and fertilization takes place on the prothalli. The fertilized female cells grow into new fern plants.

Alternative Host. One of the hosts of a plant pathogen, usually referred to the wild host(s) of a disease-causing organism of a cultivated plant.

Alterne. A sudden change in a plant species which is spread over a wide area. Such a change is brought about by variation in the soil or other environmental conditions.

Althaea (marsh mallow, hollyhock, etc.; family Malvaceae). Genus of herbs with showy, regular, five-petalled flowers and an epicalyx of 6—9 segments joined below into a cup-like involucre hiding the calyx. Several species (*e.g.* A. rosea, hollyhock) are cultivated for their flowers or (*e.g.* A. officinalis, marsh mallow) for the mucilage used in confectionery or medicine. There are about 15 species occurring in temperate Eurasia, and in N. Africa.

Altingia (family Hamamelidaceae). Genus of trees which produce valuable timber. The buds are scaly; the flowers unisexual, very reduced, and borne in small heads. The fruits form as rounded, woody heads, and are dehiscent, containing many winged seeds. There are seven species, occurring from Assam and S. China to Java.

Altricial. *ornith.* Designating birds whose young are hatched in an immature and helpless condition and remain in the nest for a period of time.

Altruism. Behaviour by an animal that decreases its chances of survival or reproduction while increasing those of another member of the same species. For example, a lapuing puts itself at risk by luring a predator away from the nest through feigning injury, but by so doing saves its offspring. Altruism in its biological sense does not imply any conscious benevolence on the part of the performer. Altruism can evolve through kin selection. if the recipients of altruistic acts tend on average to be more closely related to the altruist than the population as a whole.

Alucitidae (many-plumed moths; order Lepidoptera). Small family of fairly small months in which the forewing is divided into six plumes and the hindwing into six or seven. They are not closely related to the Pterophoridae, which are also plumed. The larvae burrow into shoots, buds, and flowers, and may induce gall-formation. They have a worldwide distribution.

Alula

1. In birds, the spurious or bastard wing; the feathered thumb.
2. In insects, a calypter.

Alveolar Ducts. The ducts which arise from each respiratiry bronchiole in mammals and terminate in air sacs whose walls get outfolded to form alveoli. (Also alveoli).

Alveolar Glands. Refer to multicellular glands (acinous glands) which have flask-shaped bodies and their necks serve as ducts *e.g.* mucous and poison gland of frogs and toads.

Alveolar Membranes. Tooth in mammals is kept in bony socket termed as alveolus and part of tooth inside the body socket is termed as root. The root of the tooth separates from bone of eveolus by collagenous periosteum of jaw bone which is as alveolar membrane (also called periodontal membrane).

Alveolate. Pitted and having the appearance of a honeycomb.

Alveoli

1. Minute air-sacs making up the lungs of a vertebrate.
2. Acini or secretory cavities of a gland.
3. Any tiny pit or cavity.

Alveolar. Of, pertaining to, or composed of alveoli.

Alveolus (plural alveoli). The name given to the millions of minute air-sacs that make up the lung. It is across the thin walls of the alveoli that the exchange of oxygen and carbon dioxide occurs, between blood and air, during respiration. The term is also used to describe the socket into which teeth fit.

Alytes Obstetricans (midwife toad; order Anura (Salientia), family Discoglossidae). Species of toad which lives in hilly terrain, hinding in crevices and under stones. It has a squat grey body with a whitish belly, and is up to 5.7 cm long. It has no vocal sacs. The male winds a band of eggs round his back legs while mating and carries them until hatching occurs. The toad is found in western and central Europe.

Amandava (waxbills, munias; family Estrildidae). Genus of small, red-billed, finch-like birds, commonly kept in captivity. They inhabit tall grass and reeds, where they build untidy nests.. A amandava (red munia) is the only estrildid with a dull non-breeding plumage.

There are three species found in India, SE Asia, and Africa.

Amanita. A white-spored fungus of the genus Amanita, as A. muscaria, the fly agaric or "Death Angel," and A. phalloides, both hightly poisonous mushrooms.

Amanitaceae (order Agaricales). Family of fungi in which the fruit bodies are mushroom-like; the gills are not attached to the stipe, the stipe can be broken cleanly from the pileus, and both universal and partial veils are typically present. There are two genera. The are found on the ground, chiefly in woodland.

Amaranthaceae (cockscombs, celosias, etc). Family of mostly annual herbs in which the leaves are entire, and without stipules. The flowers are small, borne in dense, spike-like cymes, and mostly unisexual; they have 3—5 scarious, green, or brightly coloured perianth segments; 3–5 stamens; and a one-celled ovary which forms a small, dry fruit. Celosia cristata (cockscomb) is cultivated for its red flowers. There are about 40 genera, with 500 species, mostly tropical or warm-temperate, but occurring as weeds in cooler regions.

Amaranth. A family of warm-climate plants whose flowers keep their colour even when dried. Love-lies-bleeding is a popular species.

Amaryllidaceae (daffodils, narcissi, etc.). Family of bulbous herbs which differ from Liliaceae only in the inferior ovary in their flowers. The flowers are solitary, or occur in umbels. The inflorescence is enclosed at first in a scarious spathe. The flowers are regular, and showy, with six similar perianth segments in two whorls, and sometimes with a trumpet-like corona. There are six stamens. Many species are cultivated for their flowers. There are about 60 genera, with 500 species, mainly of warm-temperate regions but extending into the tropical and cool-temperate zones.

Amaryllis. The belladonna lily, a bulbous plant from southern Africa which is cultivated for its red fragrant flowers.

The name amaryllis is also used by horticulturists for related plants whose thick stalks bear showy clusters of lilylike flowers. The amaryllis has given its name to the narcissus and snowdrop family, the Amaryllidaceae.

Amaurotic. Pertaining to or affected by blindness.

Amazilia (hummingbirds; family Trochilidae). Largest genus of hummingbirds, comprising birds that are medium-sized, with bronze-green upperparts and rufous or white underparts. They are nectar feeders, but will take insects. They breed all year round in tropical regions. There are about 31 species, found in S. and C. America, with two species reaching N. America.

Amazon

1. A slave-making ant of the genus Polyergus.
2. A South American parrot of the genus Amazona.

Amazona (Amazon parrots; family Psittacidae). Genus of medium-sized, short-tailed, mainly green parrots They feed on fruit and seeds, are hole-nesting, and are found in the forests of C. and S. America and the W. Indies. Many W. Indian species are endangered A. vittata (Puerto Rican parrot) is nearly extinct. There are 27 species.

Amber. A yellow or reddish-brown fossh resin. The resin was exuded by certain trees and other plants and often contains preserved insects, flowers, or leaves that were trapped by its sticky surface before the resin hardened. Amber is used for jewellery and ornaments. It also has the property of acquiring an electrical charge when rubbed (the term electricity is derived from electron, the Greek name for amber). It occurs throughout the world in rock strata from the Cretaceous to the Pleistocene, but most commonly in Cretaceous and Tertiary rocks.

Ambergris. A concretion from the intestinal tract of the sperm whale found floating in tropical waters, used in perfumery for the fixing of odors.

Ambidextrous. Using both hands equally well.

Ambient. The surrounding/prevailing conditions *e.g.* ambient temperature, ambient pressure etc. to which the living organisms are put.

Ambilateral. Pertaining to or affecting both sides.

Amblycera (class (Insecta, order Phthiraptera). Suborder of chewing lice, characterized by the pedunculate first flagellomere of the antenna, the groove of the head which conceals the antenna when at rest, the horizontal articulation of the mandibles, and the presence of maxillary palps. Three of the constituent families parasitize only mammals, three only birds, and one parasitizes members of both classes. The lice eat feathers, skin, sebaceous exudates, or blood. Species in one genus, Trochiloecetes, have the mouthparts modified as piercing stylets (analogous to those of Anoplura) and feed on the blood of their hummingbird hosts. There are 75 genera, with about 850 species.

Amblyopsidae (cavefish; class Osteichthyes, sub-class Actinopterygii, order Percopsiformes). Family of small freshwater fish, (*e.g.* Amblyopsis rosae of the eastern USA) of interest mainly because most species are typical cave dwellers, lacking functional eyes and body pigment. There are about five species.

Amblypoda. Term sometimes used synonymously with 'Pantodonta', but otherwise describing an order of large, archaic, and predominantly American ungulates comprising the four distinct suborders Pantodonta, Dinocerata, Xenungulata, and Pyrotheria. The Pantodonta cullminated in the hippopotamus-like Coryphodon; the Dinocerata in Uintatherium, a grotesque animal resembling a rhinoceros in size and general appearance; the Xenungulata produced animals of similar size and build; and the Pyrotheria developed forms remarkably similar to early elephants.

Amblypygi (tailless whipscorpions, whipspiders; class Arachnida). Order of dark-coloured arachnids, in which the abdomen is flat, oval, without a telson, and lacking terminal flagellum, spray glands, and spinnerets. The first pair of legs is long, whip-like, and with many false segments; they are highly sensitive to touch and are used in water location and mating. The raptorial pedipalps are held parallel to the ground and their last two segments may be almost chelate. The cheliceral are subchelate. Amblypygids resemble spiders in many ways except for the long, whip-like first pair of walking legs, strong pedipalpi, and lack of spinnerets. Arthropod prey, *e.g.* cockroaches, crickets, beetles, and spiders, are impaled on the pedipalpal spines, pre-oral liquefaction presumably taking place. Amblypygids live in humid conditions under stones, logs, bark, and litter, and some are cavernicolous. The 60 or so species are divided between two families, and are tropical and subtropical in distribution.

Amblyrhynchus Cristatus (marine iguana; order Squamata, suborder Sauria (Lacertilia), family Iguanidae). The only extant marine lizard. It feeds on seaweeds along the coasts of the Galapagos Islands, and basks on rocks or hides in fissures when not swimming. It attains lengths of up to 1.4 m and can weigh 10 kg.

Ambulacral Area. One of five radially disposed sections of plates in echinoid echinoderms through which tube feet extend.

Ambulacral Groove. A groove on the ventral surface of each ray of a sea star which contains the podia or tube feet.

Ambulacrum. Refers to groove lined with tube feet in certain echiniderms.

Ambulatory. Capable of walking or moving from place to place.

Ambystoma Mexicanum (axolotl; order Urodela (Caudata), family Ambystomatidae). Species of neotenous salamander, normally breeding as an aquatic larva, but capable of metamorphosis into a terrestrial form. It is sexually dimorphic, the female being the larger, reaching up to 29 cm in length. Normally it is black but albinos are quite common. It is confined to certain lakes, around Mexico City.

Ambystomatidae (mole salamanders, axolotls; order Urodela (Caudata), class Amphibia). Family of amphibians which have a broad head; the tongue free only at the sides; palatal teeth; costal grooves down the body; short, strong limbs; and a laterally compressed tail. Most burrow in earth for much of the year, but neotony is common in some species. There are more than 30 species, found in N. and C. America.

Ameba or **Amoeba.** A simple unicellular or acellular organism found in fresh or marine water or living parasitically in animals. Each is a naked bit of protoplasm moving by means of pseudopodia and reproducing by simple fission. Common genera are Amoeba and Entamoeba of the class or superclass Sarcodina.

Amebiasis. Infection by parasitic amebae, especially Entamoeba histolytica.

Amebocyte. A primitive, connective tissue cell capable of independent amebalike movement present in the tissues or body fluids of various invertebrates; any ameboid cell.

Ameboid. Resembling an ameba in form, structure, or movement.

Amebula. An amebalike stage in the life cycle of various parasitic protozoans.

Amelanchier (June berries; family Rosaceae). Genus of non-thorny, deciduous, small trees or shrubs with simple leaves, and flowers that are pentamerous, with 2—5 carples, united to the receptacle below but free from it above, each carpel divided into two cells by a false septum. The flowers are also in racemes, distinguishing the genus from Crataegus and Cotoneaster. Several species are cultivated for their flowers, often becoming naturalized. There are about 25 species, occurring in the north temperate zone—most N. American, but some in Europe.

Ameloblast. An enamel-forming cell in the enamel organ of a developing tooth.

Amennorrhea. Used for the absence of menstrual periods in human female; it is of two types—primary, when menstrual bleeding has never taken place and secondary if there occurs cessation of cycles in woman with previously normal periods.

Amensalism. Any association between two organisms in which one is harmed by the activities of the other. It includes parasitism and also the various ways in which plants protect themselves from pests and predators. Associations between microorganisms in which one is adveasely affected are termed antibiosis.

Ament or Amentum. A catkin.

Amentaceous. Bearing catkins.

Amentiferae. Hamamelidae.

Ameristic. Not segmented.

Amentabolous. Without metamorphosis, with reference to development of insects.

Amictic. Designating a mode of development in rotifers in which females produce thin-shelled, diploid ova which develop by parthenogenesis into amictic females.

Amino-acid. An organic substance containing carbon, hydrogen, oxygen and nitrogen (and in a few cases sulphur). Various combinations of the 25 or so amino-acids make up all the known proteins.

Amiidae (bowfin; class Osteichthyes, subclass Actinopterygii, order Amiiformes) A fairly large (up to 6 kg) fish, Amia calva, is the only living representative of its family. It is characterized by a long dorsal fin with only soft rays, a gular plate, a heuterocercal tail, and long-like airbladder. It is a carnivorous species, in which the male guards the nest and the larvae.

Amiiformes (class Osteichthyes, subclass Actinopterygii). Order of fairly primitive bony fish consisting of several families dating back to the Triassic Period, and considered by some to be intermediate between the Chondrostei and the Teleostei. Until recently they were grouped with the order Semionotiformes in the infraclass Holostei.

Amino Acid. Any of a group of water-soluble organic compounds that possess both a carboxyl (—COOH) and an amino (—NH_2) group attached to the α-carbon atom. Amino acids can be represented by the general formula R—CH(NH)$_2$COOH. R may be hydrogen or an organic group and determines the properties of any particular amino acid. Through the formation of peptide bonds, amino acids join together to form short chains (peptides) or much longer chains (polypeptides). Proteins are composed of various proportions of about 20 commonly occurring amino acids (table). The sequence of these amino acids in the protein polypeptides determines the shape, properties, and hence biological role of the protein. Some amino acids that never occur in proteins are nevertheless important, *e.g.* ornithine and citrulline, which are intermediates in the urea cycle.

Plants and many microorganisms can synthesize amino acids from simple inorganic compounds, but animals rely on adequate supplies in their diet. The essential amino acids must the present in the diet whereas others can be manufactured from them.

Structures of the 20 amino acids commonly found structure is given, the remainder of the

amino acid	symbol	R group
alanine	ala	$-CH_3$
valine	val	$-CH\langle^{CH_3}_{CH_3}$
leucine	leu	$-CH_2-CH\langle^{CH_3}_{CH_3}$
isoleucine	ile	$-CH(CH_3)-CH_2-CH_3$
proline	pro	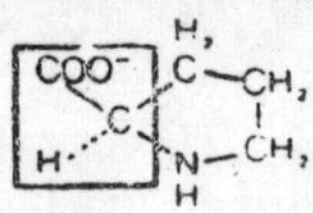
methionine	met	$-CH_2-CH_2-S-CH_3$
phenylalanine	phe	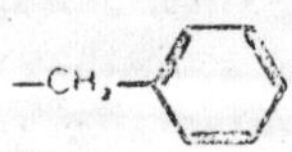
tryptophan	trp	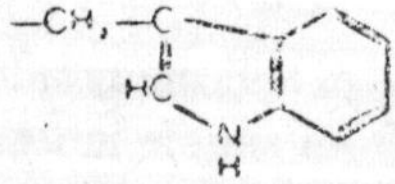

Amino acids with nonpolar R groups

amino acid	symbol	R group
glycine	gly	$-H$
serine	ser	$-CH_2OH$
threonine	thr	$-\overset{OH}{\underset{H}{C}}-CH_3$
cysteine	cys	$-CH_2-SH$
asparagine	asn	$-CH_2-C\overset{NH_2}{\lessgtr}_{O}$
glutamine	gln	$-CH_2-CH_2-C\overset{NH_2}{\lessgtr}_{O}$
tyrosine	tyr	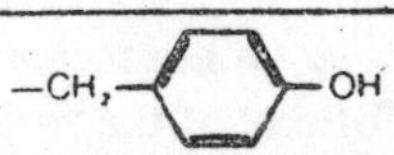

Amino acids with uncharged polar R groups

amino acid	symbol	R group	
aspartic acid	asp	$-CH_2-C\overset{O^-}{\lessgtr}_{O}$	acidic amino acids
glutamic acid	glu	$-CH_2-CH_2-C\overset{O^-}{\lessgtr}_{O}$	acidic amino acids
lysine	lys	$-CH_2-CH_2-CH_2-CH_2-\overset{+}{N}H_3$	basic amino acids
arginine	arg	$-CH_2-CH_2-CH_2-NH-\underset{\overset{+}{N}H_2}{C}-NH_2$	basic amino acids
histidine	his	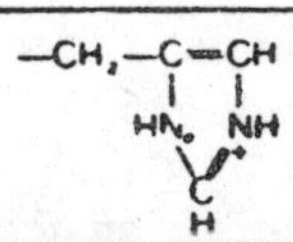	basic amino acids

Amino acids with charged polar R groups

in proteins. In all except proline, where the complete molecule has the structure

$$HOOC-\overset{H}{\underset{NH_2}{C}}-$$

Amino Acid Sequencing. The determination of the amino acid sequence of a protein. The sequencing of homologous proteins from different species has been used as a method of determining phylogenetic relationships. It is assumed that the number of differences in the sequence may be related to the length of time since the different species diverged from a common ancestor. Phylogenetic trees have been constructed from data on the sequences of cytochrome *c* and plastocyanin. These do not always correspond with information derived from other sources.

Aminopeptidase. An exopeptidase attacking terminal bond at the free amino end of the peptide chain liberating amino acids and dipeptides. They contain Mg^{2+} or Mn^{2+}.

Amino Sugar. A monosaccharide with an amino (NH_2) group in place of one of the hydroxyl (OH) groups. The most commonly occurring amino sugars are glucosamine and galactosamine. Amino sugars are components of glycoproteins.

Amitosis. Nuclear division which is characterized by the absence of a nuclear spindle and resulting in the production of daughter nuclei having unequal sets of chromosomes. The ordered process of division, duplication of nuclear membrane, and production of a spindle as in mitosis is apparently not present. Cells produced amitotically inherit veriable numbers of choromosomes.

Amixis. Failure of interbreeding to occur between different races or species.

Ammochaeta. A bristle on the head of desert ants used in cleaning sand from the anterior legs.

Ammocoets. The larva of a cyclostome like the lamprey which exhibits many primitive features and in some respects, notably the ciliary method of feeding, resembles Amyhioxus.

Ammocoetes Larva. The larva of the sea lamprey, which develops in freshwater streams, then migrates to the sea or certain inland lakes.

Ammodytidae (sand-lance, sand-eel; class Osteichthyes, subclass Actinopterygii, order Perci-formes). Family of fish with elongated bodies and long dorsal fins. The pelvic fins are often absent, as in Ammodytes tobianus (sand-eel) of W. Europe, which is caught in considerable quantity for consumption as well as bait. Feeding on animal plankton, the sand-eels themselves form an important part of the diet of cod, salmon, etc. There are about 12 species, found in marine coastal waters.

Ammonia. NH_3, a colorless, pungent gas readily soluble in water with which it combines to form ammonium hydroxide (NH_4OH). It is formed in the oxidative deamination of amino acids.

Ammonification

1. Impregnation with ammonia.
2. The conversion of complex nitrogenous compounds into simpler compounds and ultimately ammonia by hydrolysis.

Ammonite. An extinct aquatic mollusc of the class Cephalopoda. Ammonites were abundant in the Mesozoic era (225—65 million years ago) and are commonly found as fossils in rock strata of that time, being used as index fossils for the Jurassic period. They were characterized by a coiled shell divided into many chambers, which acted as a buoyancy aid. The external suture lines on these shells increased in complexity with the advance of the group.

Ammonoidea (phylum Mollusca, class Cephalopoda). Subclass of cephalopods which generally have planispirally coiled, septate shells. Characteristically the shells are tightly coiled and planispiral, although some are coiled loosely or spirally; the protoconch is globular; the shells may be either involute or evolute. Some forms have marked ventral keels; ribs and nodes may also be present. The siphuncle is variable but mainly ventral in position. Sutures are often very complex. Cameral deposits are absent. The Ammonoidea were probably tetrabranchiate cephalopods. They constitute the largest cephalopod subclass, with 163 families including the ammonites, in which the suture lines form very complex patterns; the ceratites, in which part of the suture line is frilled; and the goniatites, with relatively simple suture lines. They range in age from Devonian to Upper Cretaceous. All members are now extinct.

Ammonotelic. Excreting ammonia as the principal product of nitrogen metabolism. Compare ureotelic, uricotelic.

Amnio-cardiac Vesicles. Two pouches of splanchnic mesoderm which get fused together to form the pericardium in the embryonic development of a vertebrate.

Amniocentesis. The taking of a sample of amniotic fluid from a pregnant woman to determine the condition of an unborn baby. A hollow needle is inserted through the woman's abdomen and wall of the uterus and the fluid drawn off. Chemical and microscopical examination of cells shed from the embryo's skin into the fluid are used to detect spina bifida, Down's syndrome, or other serious biochemical or chromosomal abnormalities.

Amnion. A membrane that encloses the embryo of reptiles, birds, and mammals within the anniotic cavity. This cavity is filled with amniotic fluid, in which the embryo is protected from desiccation and from external pressure.

Amniota. Reptiles, birds and mammals, *i.e.*, land-living verte-brates whose embryos possess an amnion, chorion and allantois.

Amniote. A vertebrate whose embryos are totally enclosed in a fluid-filled sac—the amnion. Amniotes consist of the reptiles, and mammals. Compare anaminiote.

Amoe-. For words beginning with Amoe not found here.

Amoeba. A genus of Protozoa, members of which have temporary body projections called pseudopodia. These are used for locomotion and feeding and result in a constantly changing body shape. Most species are free-living in soil, mud, or water, where they feed on smaller protozoans and single-celled plants, but a few are parasitic. The best known species is the much studied A. proteus.

Amoebida (class Rhizopodea). Order of protozoa in which each cell contains a single nucleous and is of indeterminate shape. The cells form lobopodia which serve for locomotion and feeding. Most species are free-living in freshwater, moist-soil, or marine environments; a few are parasitic. There are several families, and many genera, one of which is the genus Amoeba, A. proteus, a popular subject for study in elementary biology courses, measures up to about 0.6 mm across, and is found in slow-moving or still freshwater habitats.

Amoebocyte. An animal cell whose location is not fixed and is therefore able to wander through the body tissues. Amoebocytes are named after their resemblance, especially in their movement, to Amoeba and they feed on foreign particles (including invading bacteria). They occur, for example, in sponges and mammalian blood (*e.g* , some leucocytes).

Amoeboid. Resembling an amoeba in form and movement. The term is used to describe certain gametes, *e.g.*, those of the Zygnemaphyceae.

Amoeboid Movement. Refers to the movement such as that of Amoeba in which some of the protoplasm temporarily changes from a gel into a sol and flows forwards in the form of pseudopodia. Such movement takes place in many animal cells, *e.g.*, the white corpuscles of the blood.

Amoebulae. Small amoebae that emerge after multiple fission within a cyst.

Amorphophallus (family Araceae). Genus of herbs and shrubs producing a corm-like rhizome, cultivated for food in some species. The plants bear few leaves, which are several times palmate, on a long erect stalk; and are sometimes huge, up to several metres across. The inflorescence is some times massive (1 m tall in A. titanium); the spathe blotched purple; and has a powerful, foetid odour. There are 100 species, native to Africa and Asia.

Amorphoscelidae (order Mantodea). The smaller of the two families of mantids, comprising insects which are small and mottled in colour, and which have raptorial forelegs devoid of spines on their external, lower sides. Many species have wingless females. Most live on rough-barked tree trunks against which they are well camouflaged. The pronotum tends to be shorter than that of members of the Mantidae. Most of the 50 or so amorphoscelid species are found in Australia.

AMP (adenosine monophosphate). A phosphorylated nucleoside consisting of the purine adenine and the sugar ribose phosphorylated in the 5′ position. AMP is involved in the regulation of glycolysis and gluconeogenesis, promoting the formation of fructose bisphosphate from fructose 6-phosphate (*i.e.*, promoting glycolysis) while inhibiting the back reaction to fructose bisphosphate.

Amphiaster. Used for describing the figure which is produced by two asters and spindle in cytoplasm in the division of nucleus during mitosis.

Ampherotoky. Production of offspring of both sexes by parthenogenetic animals; amphitoky.

Amphi-atlantic Species. Plants that are found on both sides of the N. Atlantic, along the seaboards of eastern North America and Europe. This disjunct distribution pattern is thought to be largely inherited from a single circum-North-Atlantic coastal-plain flora. Typical representatives are Carex paleacca (a sedge) and Ligusticum scoticum (Scotch lovage).

Amphibia The class of vertebrates that contains the frogs, toads, newts, and salamanders. Amphibians evolved in the Devonian period (about 370 million years ago) as the first vertebrates to occupy the land, and many of their characteristic are adaptations to terrestrial life. All adult amphibians have a passage linking the roof of the mouth with the nostrils so they may breathe air and keep the mouth closed. The moist scalelss skin is used to supplement the lungs in gas exchange. They have no diaphragm, and therefore the muscles of the mouth and pharynx provide the pumping action for breathing. Fertilization is usually external and the eggs are soft and prone to desiccation, therefore reproduction commonly occurs in water. Amphibian larvae are aquatic, having gills for respiration; they undergo metamorphosis to the adult form.

Amphibian. A cold-blooded animal belonging to the vertebrate class Amphibia which contains the frogs, newts and toads and salamanders. Some spend all their lives in water, and the others usually return to the water to breed. Most amphibians undergo metamorphosis during their development, hatching from the eggs as gillbreathing larvae called tadpoles. Most adults have lungs and scaleless moist skins through which they can breathe.

Amphibious. An organism that is capable of leading life both in water and on land *e.g.*, frog and small.

Amphiblastic. An alternative word for telolecithal; a term which is used to describe eggs which are having most of the yolk on one side (the vegetative pole), and the nucleus on the other side (animal pole).

Amphiblastula. The free-swimming larva of certain sponges with flagellated cells at one end large, nonflagellated cells at the other.

Amphibolurus Barbatus (bearded lizard, bearded dragon, Jew lizard; order Squamata, suborder Sauria (Lacertilia), family Agamidae). Sturdy, long-tailed agamid, up to 60 cm long : it has a spiny body with a pouched jaw, which when extended in display resembles a beard, which is larger in males. It lives in the scrub and desert of Australia.

Amphicarpous. Producing two kinds of fruit.

Amphichelydia. One of the most ancient groups of tortoises and turtles found as fossils of the Triassic period.

Amphichrony. Produing flowers of two different colors on the same stem.

Amphicoela. Refers to a group of frogs in which the vertebrae are amphicoelous or biconcave, *e.g.*, the New Zealand frog.

Amphicoelous. Biconcave; a term which is used to describe vertebrae in which the centrum is having a conical depression on both its anterior and its posterior face. Such vertebrae occur in most fish.

Amphicribal. Describing a concentric vascular bundle that has the phloem surrounding the xylem. Compare amphivasal.

Amphicyte. A satellite cell.

Amphid. A chemoreceptor located anteriorly in nematodes. Also called lateral organ.

Amphidiploid. Allopolyploid.

Amphigastria. The leaves that form in a row on the under-surface of the stem of a leafy liverwort. They are smaller than the leaves on the upper surface and are often only seen just below the apex, having been shed lower down.

Amphigenous. Growing entirely around or over a surface.

Amphigony. Sexual Reproduction.

Amphimixis. True sexual reproduction involving the fusion of two gametes. Compare apomixts.

Amphineura (phylum Mollusca). Class of elongate, bilaterally symmetrical, marine molluscs in which the shell, if present, consists of seven or eight overlapping, calcareous plates on the dorsal surface. The head is poorly differentiated, if at all. Some possess an effective radula. The posterior mantle cavity contains the anus and gills. A broad, flat foot occurs ventrally and is used for creeping. The anatomy is typically molluscan, with a primitive, ganglionic nervous system. All are blind, and lack tentacles. Sexes are generally separate. The class first appeared in the Upper Cambrian.

Amphineuran. A chiton; a mollusk of the class Amphineura.

Amphioxus (Branchiostoma). A genus of small marine burrowing cephalochordates, the lancelets. Amphioxus possesses fish-shaped body with a dorsal and caudal fin and segmentally arranged muscle blocks (myotomes).

Amphipathic. Refers to molecules which bear groups with different properties, *e.g.*, hydrophilic and hydrophobic groups.

Amphiphloic. Having phloem arranged on both sides of the xylem as, for example, in a solenostele.

Amphiphloic Siphonostele. Solenostele.

Amphiplatyan. Refers to flat on both the sides, *e.g.*, vertebrae of mammals in which centra and pleurocentra are both having more or less flat surfaces.

Amphipod. A graup of small, shrimp-like crustaceans found in both salt and fresh water. Most have bodies flattened from side to side and move on land by jumping. The sand hoppers seen jumping about in the seaweed on beaches are a common example.

Amphipoda (sandhoppers, beach-hoppers, beach fleas, water lice; class Crustacea). Order of mainly laterally compressed crustaceans, with the cephalothorax devoid of a carapace and the whole head fused to one thoracic segment, or occasionally to two. The first three pleopods are usually many-segmented swimmerets, the posterior ones modified for jumping and comprising a few segments. The gills are borne on the thoracopods. There are 4000 species, including such familiar forms as Talitrus (beach-hoppers) of the upper tide line, and Gammarus (water 'lice') of mainly fresh or brackish water. The dorsoventrally compressed Cyamus (whale lice) are host-specific parasites of whales.

Amphisbaenid. A group of worm-like burrowing lizards found in the warmer parts of Europe, Africa and the Americas. With legless bodies and scales arranged in rings, they closely resemble large earthworms.

Amphistome. A fluke of the suborder Amphistomata with acetabulum located at the posterior end of the body.

Amphistylic. A method of jaw suspension found in a few primitive elasmobranches in which the palatopterygoquadrate bar and hyomandibular cartilage are united to the chondrocranium.

Amphistylic Jaw Suspension. An arrangement which exists in a few sharks and many fossil fish (*e.g.*, the Cladoselachii) whereby the jaws are suspended by ligaments from the skull and by the hyomandibular bone.

Amphiterygidae (damselflies; order Odonata, suborder Zygoptera). Family of damselflies that resemble the Anisoptera (true dragonflies) in their habit of resting with their wings open. They are distinguishable from all other zygopteran families by having five or more antenodal cross veins, of which the first two are thickened and the rest are incomplete. The robust larvae are adapted for clinging to rocks in fast-flowing streams, and have huge, saccoid-shaped, caudal gills. The family is mainly tropical, and 19 extant species have been described.

Amphithecium. The outer layer of cells in the sporangium of certain bryophytes.

Amphitoky. Ampherotoky.

Amphitrichous. With one or a tuft of flagella at each pole, as in certain bacteria.

Amphitropous. *bot.* Having ovary inverted and attached at the middle.

Amphiuma. A strange amphibian of the south-eastern USA with a snake-like bndy up to one metre long and tiny legs. Amphiumas spend much of their time in underwater burrows, raising their heads above the water to breathe.

Amphiumidae (Congo eels, lamper eels; order Urodela (Caudata), class Amphibia). Family of eel-like, semi-larval salamanders with a cylindrical body up to 1 m long, rudimentary limbs and lungs, small lidless eyes, no tongue, and external gills that are lost in adults. They are nocturnal feeders on insects, molluscs, crustaceans, etc. They may leave the water in wet weather. Eggs are laid in a nest. Fertilization is external, with no spermatophore. There is a single genus, Amphiuma, and three species, occurring in freshwater swamps in the south-eastern USA.

Ampulicidae (cockroach wasps, digger wasps; order Hymenoptera, superfamily Sphecoidea). The most primitive family of sphecid hunting wasps, in which the antennae are inserted low on the face, with the sockets touching or close to the frontoclypeal suture. The pronotum has a high collar, and is long to very long and often tuberculate. The propodeum is long to very long, with the enclosed area entirely dorsal, and U-shaped to nearly triangular. The mid-tibiae have two apical spurs, and the claws have an inner tooth. The marginal cell of the forewing is acuminate apically, and apendiculate. The forewing has two or three submarginal cells, and two recurrent veins, (the first recurrent received by submarginal 1 or 2, the second received by 2 or 3). The abdomen is sessile, or with a petiole composed of the sternum and tergum. The male genitalia have a movable digitus and cuspis. There are about 160 species, mainly tropical, and the females provide their nests with stung and paralysed cockroaches.

Amphivasal. Describing a concentric vascular bundle in which the xylem surrounds the phloem. Compare amphicribal.

Amphixylic (amphivasal, perixylic). Used for a centric vascular bundle in which the tissues are arranged concentrically and an outer ring of xylen completely covers a central core of phloem cells. This arrangement can be seen in lily of the valley.

Amphoteric

1. Possessing opposite traits or characters.
2. Capable of acting both as a base and an acid.

Amplectant. Clasping or embracing, said of tendrils.

Amplexary Pads. Used for describing sexual dimorphic character present in hand of male frog which forms in the first digit grows and develops during breeding season under hormonal influence and serves to hold the female during amplexus.

Amplexicaul. Describing a sessile leaf in which the base of the lamina clasps the stem at the node as, for example, the upper leaves of henbit (Lamium amplexicaule).

Amplexus. Sexual embrace without true intercourse occurring, as in frogs.

Ampulla

1. *zool.* A membranous sac or vesicle or a dilatation of a tube.
2. *bot.* A bladderlike sac or vesicle, as that attached to the leaves or roots of certain aquatic plants.

Ampullae of Ear. Refers to the dilatations of the three semicircular canals of the vertebrate ear at the point where each is entering the sacculus.

Ampullae of Lorenzini. Refers to the small pits in the skin which are situated close to the lateral line canals or in patterns on the heads of many fish.

Amygdalus (almond, peach: family Rosaceae). Genus of a few species, apparently from temperate Asia, but almond, peach, apricot, and nectarine have long been cultivated and are grown in many parts of the world. They are close to Prunus, but have axillary buds in threes, and fruits that are usually downy, with the stone pitted or wrinkled.

Amylase. The hydrolase enzyme that catalyses the hydrolysis of the α(1—4) glucosidic bond in starch. Amylase occurs in two forms, designated α and β. β-amylase only attacks the nonreducing ends of a starch molecule, successively hydrolysing alternate α(1—4) linkages and releasing maltose molecules. β-amylase is found in germinating seeds and is important for the production of malt in the brewing industry. Amylose is completely degraded to maltose by β-amylase cannot attack α(1—4) linkages beyond the first branch on each chain. A separate enzyme, α(1—6) glucosidase, exists to hydrolyse the α(1—6) branching linkages of amylopectin.

α-amylase differs from β-amylase in that it can attack α(1—4) bonds within the starch molecule. It thus degrades amylopectin more completely than β-amylase.

Amyloid. A complex protein deposited in the tissues as a result of degenerative processes, resembling starch in that it stains blue-black upon the addition of iodine.

Amylopectin. A polysaccharide that, with amylose, makes up starch. It consists of glucose units linked by α(1—4) glucosidic bonds with branches formed by α(1—6) bonds. The molecular weight of amylopectin may be as high as 10^7. It can form either colloidal or micellar solutions with water.

Amyloplastid. A plastid which stores starch.

Amylopsin. A pancreatic enzyme which catalyses the hydrolysis of starch to dextrins and finally to maltose.

Amylose. A polysaccharide consisting of linear chains of between 100 and 1000 linked glucose molecules. Amylose is a constituent of starch. In water, amylose reacts with iodine to give a characteristic blue colour.

Amynodontidae (order Perissodactyla, suborder Ceratomorpha, superfamily Rhinocerotoidea). One of two extinct families of rhiaoceros-like animals, which lived during the Eocene and Oligocene in America and Eurasia, and which persisted into the Miocene in Asia. They were analogous in size and shape to the hippopotamuses and, like them, probably lived in rivers.

Amyotonia. Lack of muscle tone.

Anabaena (order Nostocales). Genus of filamentous cyanobacteria in which the filaments when viewed under the microscope resemble strings of beads. Thick-walled, clear heterocysts can be seen at intervals along the filaments. Akinetes are formed. Species are capable of fixing atmospheric nitrogen. They are common in aquatic environments, where they are planktonic, buoyed up by gas vacuoles. A. azollae lives in symbiotic association with the small floating fern, Azolla.

Anabantidae (climbing gouramis, labyrinthfish; class Osteichthyes, subclass Actinopterygii, order Perciformes). Family of fish capable of aerial respiration because of the presence of a suprabranchial organ or an auxiliary breathing apparatus. There are about 40 species, the best known of which is Anabas testudineus (climbing gourami or walking fish) of S. Asia. It lives in freshwater ponds and swamps, but can crawl over land in search of a new source of water during the dry season. The colourful gouramis, popular aquarium fish, are now placed in the family Belontiidae.

Anabiosis. A state of suspended animation during which organisms are capable of withstanding extreme desiccation and low temperatures as exhibited by moss-inhabiting rotifers and tardigrades.

Anabolic Steroid. Any steroid compound that promotes tissue growth, especially of muscles. Naturally occurring anabolic steroids include the male sex hormones (androgens). Synthetic forms of these are used medically to help weight gain after debilitating diseascs; their use by athletes to build up body muscles can cause liver damage and is banned by most athletic authorities.

Anabolism. The metabolic synthesis of complex molecules from simpler ones. It requires an input of chemical energy, which is provided by ATP. The Calvin cycle is an example of an anabolic pathway. Compare catabolism.

Anacardiaceae. Family of resinous trees, shrubs, and climbers, in which the leaves are alternate, mostly pinnate, and exstipulate. The flowers are regular, and mostly bisexual; they have five fused sepals; five free petals; 5-10 stamens; a disc; and a superior ovary, which is one- to five-locular, each locule having a pendulous ovule. The fruit is usually a drupe. There are about 77 genera, with about 600 species, which are found throughout the tropics; and a few northern outliers.

Anacardium (family Anacardiaceal). Genus of trees, with simple leaves. The fruit is a kidney-shaped nut with a resinous, irritant pericarp, surmounting a swollen, fleshy, edible peduncle. A. occidentale (cashew nut) is a bushy tree, extensively cultivated. There are 15 species, of neotropical distribution.

Anaconda. The largest snake in the world, with lengths often exceeding 7 metres. The anaconda lives in South America and spends a lot of time in the water. It is often called the water boa. It feeds mainly on birds, deer and rodents. It is not a poisonous snake, but kills by twining around and suffocating its prey.

Anacrogynous. Not possessing terminal archegonia. Compare acrogynous.

Anadromous. Ascending rivers to spawn, as salmon, shad, striped bass, and some trout. Compare catadromous.

Anaemia. Refers to the condition in which there occurs the law percentage of haemoglabin in blood than normal (16 g/dl in male and 14 g/dl in female); may be occurring due to red blood cell volume deficiency (microcytic anaemia) or with respect to haemoglobin concentration (normochromic or hypochromic). Vitamin B_{12} plays vital role in curing perniceous or microcytic anaemia. The normochromic anaemia is cured by iron therapy.

Anaerobe. An organism that can live in the absence of free oxygen. Obligate or strict anaerobes cannot live in the presence of free oxygen. Facultative or indifferent anaerobes can grow in the presence of oxygen but do not use it. Such organisms include the denitrifying bacteria and lactic acid bacteria. The term facultative anaerobe may also be used of such organisms as yeasts that can grow under anaerobic conditions but given free oxygen will use it to oxidize the products of anaerobic respiration. They thus grow better with free oxygen. However yeasts and similar organisms could equally well be described as facultative aerobes. Compare aerobe.

Anaerobic. Capable of growing in the absence of molecular oxygen.

Anaerobic Respiration. A type of respiration in which foodstuffs (usually carbohydrates) are partially oxidized, with the release of chemical energy, in a process not involving atmospheric oxygen. Since the substrate is never completely oxidized the energy yield of this type of respiration is lower than that of aerobic respiration. It occurs in some yeasts and bacteria and in muscle tissue when oxygen is absent (oxygen debt). Obligate anaerobes are organisms that cannot use free oxygen for respiration; facultative anaerobes are normally aerobic but can respire anaerobically during periods of oxygen shortage. Alcoholic fermentation is a type of anaerobic respiration in which one of the end products is ethahol.

Anagenesis. The degree or rate of evolutionary divergence.

Anal. Of, pertaining to, or located near the anus.

Analgesic. An agent which relieves pain without altering consciousness.

Analogous. Describing organs, often similar in appearance, that carry out similar functions but have different origins. For example, phyllodes are analogous to leaf blades but are derived from petioles. Compare homologous.

Analogous Organs. Organs, which, although they may have quite different origins, carry out the same functions in different animals, *e,g.*, the kidneys of an amphibian and those of a mammal (*cf.* Homologous Organs).

Anal Plate. A thin sheet of skin which is formed by fusion of the wall of the hindgut with the ectoderm in the vertebrate embryo. It marks the position where the future anus will be formed by perforation.

Anal Respiration. Refers to the absorption of dissolved oxygen which takes place through the wall of the hind-gut in certain arthropods, *e.g.*, Cyclops. To facilitate the process the water is kept moving by rhythmic contractions of the intestine.

Anal Style. A pair of small structures that are attached to sternite of 9th abdominal segment in most insects; the characteristic feature and identification marks of male cockroach.

Anal Vesicles. Refers to the chief excretory organs of echinoids; a pair of caeca which are leading into the hind-gut, each caecum having numerous ciliated funnels opening into it from the coelom.

Anamnia (Anamniota). Vertebrates whose embryos lack an amnion, *i e.*, Agnatha, Fish and Amphibia.

Anamniote. A vertebrate that lacks an amnion and whose embryos and larvae must therefore develop in water.

Anamniotes comprise the agnathans, fishes, and amphibians. Compare amniote.

Anamorpha. A subclass of Chilopoda (centipedes) which is having fifteen pairs of legs; comprising the two orders Lithobiomorpha and Scutigeromorpha.

Anamorphic. Designating a type of development in which there are only gradual changes from one stage to the next.

Anamorphosis (anamorphic growth). Developmental phenomenon in some Protura and Chilopoda (centipedes), *e.g.*, Lithobius, Bothropolys, and Scutigera; the young do not possess the full, adult complement of body segments and legs on hatching; these are developed successively in later stages, after moulting. Centipedes that hatch with only part of the adult complement of segments take several years to reach sexual maturity. Compare epimorphosis.

Anandrous. Having no stamens, said of pistillate flowers.

Anaperturate. Describing a pollen grain without any type of aperture.

Anaphase. The stage following metaphase in nuclear division, during which separation of either chromatids or homologous chromosomes commences. In mitosis and in the second division of meiosis daughter chromatids are seen to move apart towards opposite poles of the spindle. Although the nature of the forces operating to initiate this separation is as yet unknown, two processes within the spindle, both probably involving microtubule activity, are believed to contribute to chromatid movement. The first is the elongation of the spindle fibres in the equatorial region between the chromatids. The spindle eventually doubles in length. The second is the shortening of the fibres attached to the kinetochores in the centromeric regions. This is thought to be due to depolymerization of tubulin in the constituent microtubules at the ends nearest to the poles.

In anaphase of the first division of meiosis, the homologous chromosomes of each bivalent, each with a complete centromere, become separated and move towards opposite poles of the spindle.

Anaphylaxis. An abnormal immune response that occurs when an individual previously exposed to a particular antigen is re-exposed to the same antigen. Anaphylaxis may follow an insect bite or the injection of a drug (such as penicillin). It is caused by the release of histamine and similar substances and may produce a localized reaction or a more generalized and severe one, with difficulty in breathing, pallor, or drop in blood pressure, unconsciousness, and possibly heart failure and death.

Anaplasia. Reversion of cells to an embryonic type, as may occur in certain types of tumors.

Anapophyses. A pair of small prominences below the post-zygapophyses on certain vertebrae, *e.g.*, on the lumbar vertebrae of mammals.

Anapolytic. Designating tapeworms which fail to shed the terminal proglottid when it is ripe.

Anapsid. A type of skull having a complete covering of dermal bones with no temporal fossae; such a skull is present in the primitive reptile Seymouria and also in the Chelonia (tortoises and turtles).

Anapsida. Reptiles in which the skull is having a complete covering of dermal bones with no temporal fossae, although the bones may be emarginated from the hind-end. Examples are the fossil Cotylosauria as well as tortoises and turtles.

Anaspida. Used for describing ostracoderm fish of the Silurian and Devonian periods that has a laterally compressed fusiform body, a heterocercal tail and long thin rectangular scales arranged in parallel rows on the body.

Anastomosis

1. The union of a number of structures as blood vessels, nerves, or veins of a leaf so that a branching network is formed.
2. The joining by surgery of two hollow structures.

Anapromorphidae (order Primates, suborder Prosimii, infraorder Tarsiiformes). Extinct family of tarsier like animals showing signs of tarsioid specialization. The brain was similar to that of a modern tarsier but the olfactory regions were more highly developed, and the temporal and occipital lobes enlarged. The eyes were large, the face short. There were at least 20 genera, distributed in Europe and America from the Palaeocene to Oligocene.

Anarhichadidae (wolffish; class Osteichthyes, subclass Actinopterygii, order Perciformes). Family of marine fish which are eel-like in shape, have a very long dorsal fin, lack pelvic fins, but have powerful jaws with strong teeth used to crush hard-shelled molluscs and crustaceans. The common wolffish, Anarhichas lupus, is frequently caught in trawls and marketed fresh or as frozen fillets. There are about six species, living in cold to Arctic waters.

Anas (dabbling ducks; family Anatidae). Genus of mainly fresh-water, short-legged, narrow-winged ducks, with typically flattened bills. Almost all have a speculum. They feed mainly by dabbling of up-ending on vegetable matter and invertebrates. There are 36 species, with cosmopolitan distribution, most species being migratory. They are typified by A. platyrhynchos (mallard) which is the ancestor of most domestic ducks.

Anastral. Lacking an aster, with reference to the achromatic figure.

Anaspidiformes (class Cephalaspidomorphi). Extinct order of fossil fish-like vertebrates considered to be related to the lampreys. They had a fussiform body with a hypocercal tail, and up to 15 pairs of external gill openings. Extant during the Silurian and Devonian Periods; these small (up to 15 cm in length) animals lived predominantiy in fresh water.

Anatase (octahedrite). Mineral, specific gravity 3.9, hardness 5.5-6.0, chemical composition TiO_2. It occurs as an accessory mineral in metamorphic and igneous rocks, and is also found is hydrothermal veins and granite pegmatites. Its most distinguishing feature is its octahedral shape. In colour it may be yellow, shades of brown, blue, or black, with a white streak, and it has a vitreous lustre. It crystallizes in the tetragonal system. The crysfals are normally bipyramidal octahedra, but the mineral can also be tabular in shape.

Anatidae (ducks, geese, swans; order Anseriformes). Family of mainly aquatic birds with flat, lamellate bills (except Mergus). The front toes are webbed. Many Anatidae are sexually dimorphic They have thick feathers with insulating down. The flight feathers are moulted simultaneousiy after breeding. They feed on vegetable and animal foods. They nest on the ground or in holes in trees, among rocks, or in the earth, and the nest is usually lined with down. There are 143 species, with cosmopolitan distribution.

Anatomy

1. The science which deals with the structure of animals or plants.
2. The structural organization or makeup of an organism, or a treatise dealing with the same.

Anatoxin. A toxin modified by heat or chemical treatment so that its toxic properties are lost but its antigenic properties retained.

Anatropous. Describing the form of ovule orientation in the ovary in which the funiculus has lengthened and the ovule turned through 180° so that the micropyle is folded over and lies near the base of the funiculus (illustration at ovule). This arrangement is the most common. Compare campylotropous, orthotropous.

Anax Imperator (emperor dragonfly; order Odonata, family Aeshnidae). One of the largest dragon-flies occurring in Eurasia. The thorax is pale jade-green; the abdomen in males is mainly blue, with fine, black markings—in females the blue is replaced with olive-green. The females hawk high in the air in woodland areas and return to ponds or lakes only to mate and lay eggs. The males rarely leave the vicinity of water, and remain on the wing for long periods.

Ancalomicrobium. Genus of Gram-negative bacteria not assigned to any taxonomic family. The cells are variable in shape, according to conditions, but typically have 2-8 stalk-like extensions. Reproduction is by budding from one site on the cell body (not from stalks). These bacteria can grow under aerobic or anaerobic conditions, and are chemo-organotrophs. They are found in aquatic environments.

Ancestrula. In bryozoans, the first zooid which, by budding, gives rise to daughter zooecia.

Anchor. A curved hook in the opisthaptor of a monogenetic fluke.

Anchovy. A small marine fish, only 14 cm long, related to the herring. It lives in large shoals in estuaries or shallow bays, particularly in the Mediterranean, but goes to deeper water in winter. Before the Dutch Zuider Zee (now the I Jsselmeer) was dammed and partly reclaimed it was a rich source of anchovies.

Anchovy

Ancipital. Having two edges, said of flattened stems.

Ancyclostomiasis. Hookworm disease resulting from infestation with Ancylostoma duodenale

Ancistrocladaceae. Family of mostly woody climbers, with twig tips modified to form coiled hooks. The leaves are simple and alternate, with stipules absent or soon caducous. Flowers are cymose, regular, and bisexual. They have five fused sepals and petals, the former enlarging in the fruit (a nat); 10 (or occasionally five) stamens; the ovary semi-inferior, with a single basal ovule; and three connate styles. There is one genus, containing about 10 species, with a disjunct distribution in W. Africa and SE Asia.

Ancylopoda (Chalicotheres; superorder Mesaxonia, order Perissodactyla). Extinct suborder of horse-like animals, some with forelimbs longer than hind limbs, distributed widely in the Holarctic from the Eocene to Pleistocene. They differed from contemporary perissodactyls in possessing three toes whose terminal phalanges were cleft, and undoubtedly bore claws rather than hoofs. Probably they fed on roots and tubers and the claws and long forelimbs were used in digging.

Andalusian Fowl. The blue hybrid offspring of a cross between black and white-splashed domestic fowl.

Andrena Fulva (tawny mining bee; order Hymenoptera, family Andrenidae). Onc of the commonest and most conspicuous of the solitary mining bees that appear in April in Europe, including Britain. The females are clothed in a dense coat of bright, fox-red hairs. Although strictly solitary, they often nest in dense aggregations in well-drained soils, especially in lawns in town gardens. Brood cells are excavated one at a time at the end of lateral branches from the main, vertical burrow. The females forage at a wide range of spring flowers. The males are more slender than the females, and are drab brown with a tuft of silvery hairs on the face below the antennae. There is a single generation each year, and new adults overwinter in their natal cells until the following spring.

Andrenidae (solitary mining bees; order Hymenoptera). Large family of short-tongued, solitary bees. Each antennal socket is connected to the clypeus by two sutures. The tongue is pointed. The segments of the labial palps are similar, or sometimes the first or second is elongate and flattened. The coxae of the mid-leg are externally shorter than the distance from their apices to the posterior wing bases. The pollen scopa is a dense fringe of hind tibial, branched hairs, often augmented by a coxal floccus of long, curved hairs. The brood cells are lined with a wax-like secretion of the Dufour's gland.

Androconia. Specialized scales on the wings of certain male lepidopterans which serve as outlets of odoriferous glands.

Androconial Scale (androchronial scale). In male Lepidoptera, a wing or body scale that is modified for the dispersal of a sexual scent (pheromone). Glandular cells in the wing membrane or integument of the body connect with the base of the scale, but just how the scent is dispersed is not fully understood. The scent may diffuse via the hollow body of the scale to the fimbriae, or minute hairs, at the scale tip; or it may simply pass on to the surface of the scale. Androconia may be scattered singly over the surface of the wing, but are sometimes grouped together in tufts.

Androdioecious. Describing plant species in which male and hermaphrodite flowers are borne on separate individuals, as in Caltha (marsh garigolds). Compare andromonoecious, gynodioecious.

Androecium. The male component of a flower, made up of several stamens and sometimes also staminodes. The androecium usually surrounds the gynoecium, although the exact arrangement or position may not be symmetrical in the more advanced forms. The androecium is described as apostemonous when the stamens are separate, monadelphous when the filaments are all fused, and syngenesious when the anthers are fused. If the stamens form two groups, the members of each group being joined by their filaments, the androecium is diadelphous and if three or more groups are formed in this way, polydelphous. Petalostemonous describes the condition where the filaments are joined to the petals. The androecium is represented in the floral formula by the letter A.

Androgamete. A male gamete or germ cell; a spermatozoan.

Androgametophyte

1. *a.* The male gametophyte. *b.* The gametophyte which produces microgametes.
2. In seed plants, the germinated pollen grain.

Androgen. One of a group of male sex hormones that stimulate development of the testes and of male secondary sexual characteristics (such as growth of facial and pubic hair in men). Testosterone is the most important. Androgens are produced principally by the testes when stimulated with luteinzing hormone but they are also secreted in smaller amounts by the adrenal glands and the ovaries. Injections of natural or synthetic androgens are used to treat hormonal disorders of the testes and breast cancer and to build up body tissue (anabolic steroid).

Androgenesis. Development of a fertilized ovum from which the egg nucleus has been experimentally removed or inactivated. Compare pseudogamy.

Androgenous. Pertaining to the production of males or male gametes.

Androgynophore (androphore). An extension of the receptacle between the petals and the stamens on which the androecium and gynoecium are borne. It is seen in many members of the Capparaceae.

Androgynous

1. *zool.* Hermaphroditic; having characteristics of both sexex.

2. *bot.* Possessing both staminate and pistillate flowers in the same cluster.

Android. Resembling a male.

Andromonoecious. Describing plant species in which male and hermaphrodite flowers are borne separately on the same individual, as in Aesculus hippocastanum (horse chestnut). Compare androdioecious, gynomonoecious.

Androphore. A supporting structure bearing raised stamens.

Androsperm. A male-producing spermatozoan; one containing the Y-chromosome.

Androsporangium. In seed plants, a sporangium that produces androspores; a microsporangium.

Androspore

1. In seed plants, a spore that develops into a male gametophyte; a microspore or pollen grain.
2. In certain algae, a form of zoospore which gives rise to dwarf male plants as in the Oedogoniales.

Androsporophyll. A microsporophyll.

Androstane. A steroid, $C_{19}H_{32}$, the parent substance or androgenic hormones.

Androsterone. An androgenic steroid present in the urine of men and women, considered to be a catabolic product of testosterone.

Androstrobilus. A strobilus which bears microsporangia or pollen sacs.

Anecdysis. A condition in which molting ceases and no further growth occurs, as in crustaceans.

Anemia. A pathological condition characterized by a marked reduction in circulating red blood cells, or hemoglobin, or both.

Anemochorous. Dispersed by the wind.

Anemone. Several species of flowering plant related to the buttercup and often known as windflowers. The most common is the wood anemone which froms a thick white carpet in woodlands. Its leaves are arranged in a whorl beneath the solitary flower. The short-stalked yellow anemone and the blue anemone may also be found in woodlands.

Anemophilous. Pollinated by the wind.

Anemophily (wind pollination). Pollination by pollen carried on the wind. Wind-pollinated flowers often have reduced sepals and petals and often appear before the leaves. This helps ensure that the stigmas are effectively positioned for pollen interception and the stamens are free to release their pollen. The stamens often have very long filaments while the styles may be long and feathery. As female and male parts compete for room they are often found in different flowers on the same plant, as in hazels (Corylus), or on separate plants, as in black bryony (Tamus communis).

The pollen of wind-pollinated plants needs to be light and smooth surfaced and is only released from the anthers on warm dry days. In catkin-bearing species it can be stored in saucer-like bracts until disturbed and transported on a windy day. Wind-transmitted pollen has a typical diameter of 20-30 μm and may be carried thousands of miles but will normally travel less than 1 km. Because air movement is random pollen needs to be produced in vast quantities (a hazed catkin may produce 4,000,000 pollen grains). This is wasteful of the plant's resources and is thought by some to be the most primitive form of pollination. Others consider that the surviving species using this method show evidence of secondary modification from entomophilic forms. Compare entomophily, hydrophily.

Anemotaxis. A turning of an animal in response to air currents.

Anemotropism. Orientation of a plant or a part of it in response to air currents.

Anesthesia. Loss of sensation, as that produced by drugs.

Anesthetic

1. Pertaining to or causing anesthesia.
2. *zool.* A drug which reduces motility.

Anestrus. In a female mammal, the interval of sexual inactivity between periods of heat (estrus). Also called diestrum.

Aneuploidy. The condition, which is produced by nondisjunction of homologous chromosomes at meiosis, in which one or more chromosomes may be missing from or added to the normal somatic chromosome number, If both of a pair of homologous chromosomes are missing, nullisomy occurs. Monosomy and trisomy refer to the conditions in which one or there homologues occur respectively, instead of the normal two. Polysomy, which includes trisomy, refers to the state in which one or more chromosomes are represented more than twice in the cell.

Aneurism. Refers to the localised dialation of an artery wall. The elastic walls tend to balloon and if collagen fibres sheath breaks down artery then it is having tendency to undergo ballooning.

Aneurysm. A circumscribed difatation in the wall of an artery forming a soft, pulsating, blood-filled tumor.

Aneuspory. The production through a modification of the meiotic process, of an unusual number of spores (usually two) instead of the four normally formed from each spore mother cell. It is seen in the formation of megaspores in dandelions (Taraxacum) where, after the first meiotic division the chromosomes stay in the one cell forming a restitution nucleus. The second meiotic division gives rise to two cells each with an unreduced number of chromosomes. One of these develops parthenogenetically into an embryo. Crossing over and hence reassortment of the genes can occur during the first meiotic division. This accounts for some of the variation found in apomictic complexes that have arisen by aneuspory.

Angeltish. The name commonly given to three quite unrelated fishes with wing-like fins and bright colours. One is a shark (the Monkfish), one a small fish of coral reefs, and one a freshwater fish which is a great favourite for aquaria.

Angelfish

Angelica. (family Umbelliferae). Genus of tall perennialherbs with stout, hollow stems. They have two or three leaves, which are pinnate with broad oval leaflts. The umbels are compound, with many rays, few bracts, but many bracteoles. The fruits are ovoid, and flattened dorsally, with two broad, marginal rays, and three dorsal ridges on each carpel. A archangelica is cultivated and candied as a sweetmeat. There are some 70 species, occurring in the northern hemisphere, and in New Zealand.

Angelica, Wild. A tall herb belonging to the carrot family and found mainly in fens and wood. The grooved stem supports clusters of small white or pink flowers. The sweet angelica used in cooking comes from the stems of the closely related garden angelica which bears yellow-green flowers.

Angina Pectorts. A substernal pain which is developed in chest when myocardium gets ischemic (Also myocardial ischemia) during exertion due to the accumulation of p factor in muscles.

Angiogenesis. The development of blood vessels.

Angiology. The study of blood and lymphatic vessels.

Angiolysis. The atrophy or obliteration of functional blood vessels in embryonic or postnatal development.

Angioma. A tumor composed principally of blood or lymphatic vessels.

Angiosperm. A large group of plants which produce flowers and bear their seeds in fruits.

Angiospermae. The flowering plants : a subdivision of the Spermatophyta or a class of the Pteropsida. Angiosperm gametes are produced within flowers and the ovules (and the seeds into which they develop) are enclosed in a carpel (compare Gymnospermae). The angiosperms are the dominant plant forms of the present day. They show the most advanced structural organization in the plant kingdom, enabling them to inhabit a very diverse range of habitats. There are two classes (or subclasses) within this group : the Monocotyledonae with one seed leaf (cotyledon) in the seed, and the Dicotyledonae with two seed leaves.

Angiotensin. Either of two related peptide hormones that raise blood pressure. Angiotensin I is derived from a protein (angiotensinogen) secreted by the liver into the bloodstream. As blood passes through the lungs, an enzyme splits angiotensin I, forming angiotensin II. This causes constriction of blood vessels and stimulates the release of the hormones vasopressin and aldoster-one which increase blood pressure.

Angiotonin. The old name for hypertension or angiotensin II.

Angle Gland. A gland at the corner of the mouth in birds.

Anglerfish. A group of marine fishes which catch their food with a sort of 'fishing rod'. The 'rod' is a modified fin spine, and it hangs near the mouth with a worm-like flap of skin acting as the bait. Small fishes investigating the bait are quickly snapped up. The anglerfishes have very large heads and mouths. They are found in warm and temperate seas and most live on the bottom. Some species live at great depths.

Anglerifish

Angle Shade. A common European moth which at rest looks like a dead leaf. It flies all year round. The caterpillar is green or brown, with V-shaped markings on the back, and feeds on a variety of green plants.

Angle shade

Anglewing. A small butterfly of the genus Polygonia, called the "comma" because of commalike markings.

Angleworm. An earthworm used extensively as fish bait.

Angophora (family Myrtaceae, subfamily Leptospermoideae). Genus of trees and shrubs which are closely allied to eucalypts, but which have flower petals and no operculum, and opposite leaves. They are found in or near rain forests in the east. A costata (Sydney red gum or smooth-barked apple) is a tree growing to at least 14 m, with red bark and white flowers in spring and summer, It is the dominant tree of deep sands behind coastal. dunes in New South Wales, and on alluvial sands. There are eight species, all endemic to Australia.

Angora Hair. A hair which grows to a considerable length before being shed, as those of the head.

Angora Rabbit. Å long haired domestic breed raised for its fine wool.

Angstrom Symbol : A. A former unit of length equal to one thousandth of a micrometre or one tenth of a nanometre, *i.e.* 10^{-10} metre.

Anguillidae (eels; class Osteichthyes, subclass Actinopterygii, order Anguilliformes). Family of snake-like fish with long, cylindrical bodies, narrow vertical gill slits, and a long dorsal fin continuing through the tail fin into the anal fin. The skin bears minute scales. Widely distributed in tropical and temperate regions, the adults tend to live in fresh water for a number of years, finally migrating to the sea to spawn. Being a valuable food fish, eels are marketed in large numbers in many countries. The common eel, Anguilla anguilla, is known to spawn in the Sargasso Sea, W. Atlantic, with the juveniles, called elvers, taking up to three years to reach the coast of Europe. There is one genus, Anguilla, with about 15 species.

Anguilliformes (class Osteichthyes, subclass Actinopterygii). Order of eel-like fish with elongate, cylindrical bodies, narrow gill apertures, a tail-fin confluent with the low dorsal and anal fins, no ventral fins, and the skin either naked or with minute scales. The order comprises about 22 families, including the Anguillidae (eels), Muraenidae (moray eels), and Congridae (conger eels).

Angular Bone. A dermal bone in the lower jaw of certain vertebrates.

Angwantibo. A rare member of the loris family found in the forests of West Africa. The angwantibo has a furry body about 25 cm long, large eyes and relatively long limbs. There is no visible tail. The hands and feet are highly developed for grasping branches and angwantibos move nimbly through the trees.

Anhimidae (screamers; order Anseriformes). Family of black or grey, short-necked, fairly long-legged, ungainly birds. The bill is chicken-life, the toes are long and semi-palmate, the wings broad with a spur on the leading edge. There are subcutaneous air sacs. The ribs lack uncinate processes. These birds are found in marshy habitats where they wade or swim. They are vegetarian and nest on the ground near water. There are two genera, with three species, found in tropical and subtropical S. America.

Anhinga. The water turkey or snakebird.

Anhinga (darters; family Anhingidae). Genus of long-necked birds with an elongated body, long, narrow, pointed bill, and long, stiff tail. They are totipalmate. The food consists mainly of fish, caught under water, and darters often swim with only the head and neck above water. These birds are excellent fliers, and are migratory in some areas. They inhabit the wooded shores of lakes and rivers, breeding colonially in trees. There are four species, often considered to be one, found in all the major continents except Europe.

Anhydrous. Completely lacking in water.

Ani. The ains are insect-eating birds of North and South America. They are related to the cuckoos, although they do not lay their eggs in other birds nests. Their nesting behaviour is unusual, however, because they live in colonies and several pairs usually combine to build a communal nest. The birds are about 40 cm long and shiny black in colour.

Aniline Stain. Any of a group of dyes derived from aniline that are used to stain biological material. Aniline sulphate and aniline hydrochloride both stain lingin yellow. Aniline blue is often used as a counterstain with safranin.

Animal Any member of a group of livings thing distinguished from plants by the following generel characteristics : a more compact and definite body form; absence of rigid cell walls of cellulose; power of locomotion and quick response to stimuli; inability to manufacture foods from inorganic substances. Compare plant, protist.

Animal Behaviour. The activities that constitute an animal's response to its external environment. Certain categories of behaviour are seen in all animals (*e.g.* feeding, reproduction) but these activities involve different movements in different species and develop in different ways. Some movements are highly characteristic of a species (instinct) whereas others are more variable and depend on interaction between innate tendencies and learning during the individual's lifetime. Physiologists study how changes in the body (*e.g.* hormone levels) affect behaviour, psychologists study the mechanisms of learning, and ethologists study the behaviour of the whole animal : how this develops during the individual's lifetime and how it evolved through natural selection.

Animalcule. Any minute or microscopic animal; a protozoan.

Animal Kingdom. One of the two divisions of all living things, loosely defined as comprising those organisms which move about freely and feed on plants or other animals. There is no mistaking the kingdom (plant or animal) to which the more advanced animals belong. But there is a grey area among primitive creatures, particularly the single-celled protozoa, where it is difficult to define an organism as being wholly animal or plant.

Animal Pole. The region of an ovum which contains the nucleus and the more active portion of the cytoplasm; the region that gives off the polar bodies. Compare vegetal pole.

Animal Starch. Glycogen.

Animation. Pertaining to animal-like activities that are not totally natural but tend to simulate the essential features of normal activities *e.g.* living activities of frog during hibernation which resemble the normal life activities in essential features but are not exactly similar to them.

Anise. An annual herb from the Mediterranean area belonging to the carrot family. It is grown for its spicy seeds which are used to flavour sweet foods.

Anisodactyl. *ornith.* Having feet with three front toes and one hind toe, the hallux.

Anisogamous. Having gametes differing in size and structure; heterogamous.

Anisogamy The production or fusion of motile gametes that differ in size. This condition is found in various algae and fungi. Compare isogamy.

Anisomerous. Having unequal numbers of structures in floral whorls, as four petals and six stamens.

Anisophylly. Having leaves of two different types.

Anisosporous. Heterosporous.

Anisoptera (dragonflies; class Insecta, order Odonata). One of the three suborders of dragonflies, comprising insects which are generally more robust, and fly more strongly, than that in the other suborders. The anisopteran hind-wing has a broader base than the forewing, both parts being held outspread when the insect is at rest. The nymphs are aquatic and robust, with an elaborate system of tracheal gills arranged in longitudinal rows within the rectum. Adults are often brightly coloured, and fly near water. This cosmopolitan suborder has nine families, containing more than 5000 species.

Anisozygoptera (dragonflies; class Insecta, order Odonata). Smallest of the three suborders of dragonflies, comprising insects that are superficially, similar to the Anisoptera, but whose wing have narrow bases (a zygopteran feature). The nymphs also show features held by both the other suborders. The Anisozygoptera flourished during the Mesozoic, when they replaced the ancestral dragonfly order Protanisoptera. Today there are only two, poorly known, living species, found in India and Japan.

Ankle. The joint between the leg and the foot; the region of this joint.

Ankylosis

1. A stiffness or immobility of a joint.
1. The union or fusion of two separate bones.

Anlage. A primordium or the first accumulation of embryonic cells which constitutes an identifiable beginning of a structure. Compare blastema.

Annectant. Linking; connecting together.

Annelid. Any member of the phylum Annelida, segmented, coelomate animals which comprise the polychaetes, mostly marine forms as the clam worm; oligochaetes, terrestrial or freshwater forms as the earthworm; hirudineans, parasitic forms as the leech.

Annelida. Phylum of coelomate worms which possess a definite head, and good metameric segmentation. Anatomically they are more complex than the Platyhelminthes. They are vascular, respiratory, and nervous systems which are well developed. The body is elongat, and each segment has bristles. They occur in marine, freshwater, and terrestrial environments. Some are parasitic. They are represented today by the earthworms, sandworms, and leeches. Their fossils are found in rocks dating from the Cambrian, and possible fossil annelid worms are known from Precambrian sediments in S. Australia.

Anniellidae (shovel-snouted legless lizards; order Squamata, suborder Sauria (Lacertilia). Family of limbless, burrowing lizards, inhabiting sandy ground. The head is wedge-shaped, with a countersunk lower jaw, small eyes with lids, and no tympanum. The scales are smooth. The diet consists mainly of insect larvae. There are two species, both found only in California.

Annona (family Annonaceae). Genus of shrubs or small trees, producing fruits that are aggregate, individual berries being immersed in the fleshy receptacle. Many are edible, *e g.* neotropical A. muricata (soursop), A. reticulata (bullock's heart or custard apple), A. cherimola (cherimoyer), and A. squamosa (sweetsop or sugar apple). There are 120 species, occurring in warm-temperate to tropical regions, especially in America.

Annonaceae. Family of shrubs, monopodial trees, and climbers, with aromatic tissues. The twigs have septate pith. The leaves are simple, alternate, and exstipulate. The flowers are bisexual or unisexual; often fragrant; and their parts are free. They have three sepals, six petals, and many stamens. The few to many carpels are superior, and there are one to many ovules. Fruits typically are clustered, and fleshy. Asimina triloba (pawpaw), of N. America, is one of the few temperate species. There are about 120 genera, with about 2000 species, almost entirely tropical, and found mainly in the Old World, especially in lowland rain forest.

Annovulatory Cycles. Menstrual cycles in a woman during which there occurs no ovulation.

Annual. A plant that germinates from a seed, grows, flowers, produces seeds, and then dies within a single year. Examples are marigold (Calendula officinalis) and poppies (Papaver). Compare ephemeral, biennial, perennial.

Annular. Ring-shaped or forming a ring.

Annularis. The fourth finger of the hand.

Annular Ligament

1. A circular ligament, as that which encloses *a.* the head of the radius; *b.* the footplate of the stapes.
2. A fibrous membrane separating the cartilage rings of the trachea.

Annular Thickening. A type of secondary wall patterning in tracheary elements in which the secondary cell wall is laid down in rings (illustration at tracheary elements). Annular thickening is common in tracheary elements that have not yet finished elongating, for example those in protoxylem. Compare pitted thickening, reticulate thickening, scalariform thickening, spiral thickening.

Annulata. Old name for the phylum Annelida.

Annual Ring (growth ring). The annual increase in girth of the stems, or roots of woody plants, due to the cambial activity. One can, see the annual rings of plants growing in temperate climates in cross-section as two consecutive rings of light and dark coloured xylem tissue. There are formed from a zone having larger vessel elements produced by the cambium in the spring (lighter layer), followed by a zone having smaller vessel elements (darker layer) produced during the late summer. This process is repeated annually so that the number of light or dark rings gives idea about the age of that part of the plant.

Annulate

1. Composed of rings or ringlike structures.
2. An annelid.

Annulus

1. A ring or ring-shaped structure, marking, opening, or space.
2. In mushrooms, a membranous ring that surrounds the stalk after expansion of the pileus.
3. A ring of cells at the openning of the capsule in mosses or encircling the sporangium in ferns.
4. The transverse groove or girdle of a dinoflagellate.
5. A division of a somite in leeches.
6. The seminal receptacle of a female crayfish.
7. A concentric growth ring (circulus) on the scale of a fish.

Anoa. Dwart cattle standing one metre high at the shoulder and resembling antelopes in general appearance. Anoas are found in the mountains of the Celebes islands in Indonesia.

Anobiidae (furniture beetles; class Insecta, order Coleoptera) Family of small, pubescent, brown beetles, 2-7 mm long, in which the head is hooded by the pronotum. The antennae have the last three segments elongate. Larvae are white, fleshy, and C-shaped; they have tiny legs. They bore into the wood and bark of dead trees. Larvae of Anobium punctatum (woodworm) attack furniture and structural timbers; distinctive holes are caused by the emergence of adults. Xestobium rufovillosum (death-watch beetle) bores into structural timbers. Adults make a 'ticking' noise, by tapping the head against wood as a signal to the mate; this is sometimes heard by those keeping vigil with the dying and so is thought to presage death. There are 1100 species.

Anodontia. Absence of teeth.

Anodynia. Absence of pain.

Anole. A group of small slender lizards with long tails and triangular heads. They are related to the iguanas and are found only in the Americas. Most of them live in the trees and climb by means of sharp claws and adhesive pads on the toes. Anoles can change colour with their mood, from brown to green.

Anolis (anoles; order Squamata, suborder Sauria (Lacertilia), family Iguanidae). Genus of iguanid lizards, most of which are small, arboreal eaters of fruit and insects, with slender bodies, and long whip-like tails. The clawed toes have adhesive pads, some with enlargements which give a parachute effect. They are adept at colour-change. There are 165 species, distributed from N. Carolina (USA) to S. Brazil, and in the W. Indies.

Anomalodesmata (phylum Mollusca, class Pelecypoda). Subclass of equivalve, or nearly equivalve, pelecypods most of which are relatively elongate. They have an aragonitic shell, and many genera have a nacreous endostracum. The ligament is opisthodetic if present, and the musculature is nearly isomyarian. There is a thickened hinge margin, and dentition is desmodont or edentulous. Living members have eulamellibranchiate, or occasionally septibranchiate, gills. They first appeared in the Middle Ordovician. There is only one order, the Pholadomyoida.

Anomalopidae (lantern-eye fish; class Osteichthyes, subclass Actinopterygii, order Beryciformes). Family of primarily warm-water, Indo-Pacific fish possessing a cream-coloured light organ beneath each eye. The fish can turn their 'lanterns' off at will by masking the light organ with a piece of black skin. Anomalops katopteron can 'flick' its light at the rate of about four to five times a minute.

Anomaluridae (scaly-tailed squirrels; order Rodentia, suborder Sciuromorpha). Family of arboreal, squirrel-like rodents, usually brightly coloured, and in all genera but Zenkerella possessing a patagium supported laterally by an olecranon cartilage, used in gliding. The tail is bushy, with a double row of raised scales on the underside, and in some species it is long. The limbs are pentadactyl, the claws strong. The skull is high, the palate narrow, the eyes and ears large. The cheek teeth are rooted, brachyodont with four or five cusps. There are about 12 species in four genera : Anomalurus, Anomalurops, Idiurus, and Zenkerella. They are found in W. and C. Africa.

Anomaly. A deviation from normal; an abnormal or irregular structure; a malformation.

Anomocytic. Subsidiary cell.

Anopheles (malarial mosquitoes; class Insecta, order Diptera, suborder Nematocera, family Culicidae). Genus of mosquitoes which occur all over the world and which are responsible for the transmission of malaria wherever they occur. Only females feed on blood, after piercing a subcutaneous blood vessel of some vertebrates. Males feed on nectar and other plant juices. The eggs of Anopheles are generally laid singly in small bodies of water. The larvae are surface-feeders, and have feeding bristles which are relatively small compared to those of other groups of mosquitoes. Adult Anopheles can be distinguished by their resting attitude, with the body making an acute angle with the substrate as the body is held in line with the proboscis.

Anopheles Maculipennis (malarial mosquito; class Insecta, order Diptera, suborder Nematocera, family Culicidae), The most important vector of the protozoan malarial parasite in Europe. This species is part of a so-called 'species group' comprising several sibling species (species which are morphologically similar). The female A. maculipennis lays 40-100 eggs during her lifetime.

Anopheline. Of or pertaining to mosquitoes of the genus Anopheles which serve as vectors of the causative organism of malaria in man.

Anoplura (sucking lice; class Insecta, order Phthiraptera). Suborder of lice characterized by piercing mouthparts (made up of three stylets), and by the fusion of all three parts of the thorax. All species are obligate parasites of placental mammals and feed on blood, this being digested with the aid of symbiotic bacteria. Some species are vectors of disease in humans and other animals. There are 15 families, 43 genera, and about 500 species known. Approximately 840 species of mammal are known to harbour Anoplura, and more are expected to be discovered.

Anopluran. Any insect of the order Anoplura which includes the sucking lice.

Anostraca (fairy shrimps. brine shrimps; phylum Arthropoda, subphylum Crustacea, class Branchiopoda). Order of mainly freshwater, branchiopod crustaceans lacking a carapace, and with an elongate body comprising 20 or more segments, with limbs borne by segments 11-19. The brine shrimps, Artemia species, are found in salt lakes throughout the world and tolerate very high salinities. There are 175 species.

Anostracan. Any crustacean of the order Anostraca which includes the brine shrimp and fairy shrimp.

Anoxia. Refers to condition of the lack of oxygen in tissue which diving animals *e.g.* mammals and birds suffer during submergence. Its effect is first on central nervous system so the diving animals are utilising oxygen stores in the body, oxygen is utilised only by brain and heart during this period and blood flow to other parts/organs get reduced and these tissues get adapted to anaerobic pathways.

ANS. Autonomic Nervous System.

Anser (geese; family Anatidae). Genus of large, long-necked birds with strong bills used for grazing. They are vegetarian and ground-nesting A. anser (greylag goose) and A. cygnoides (swan goose) are both extensively domesticated. There are nine or ten species with a Holarctic distribution, and they are migratory.

Anserine

1. Gooselike.
2. Of or pertaining to the order Anseriformes which includes the screamers, swans, ducks, and geese.

Anseranas (magpie goose; family Anatidae). Mono-typic genus comprising A. semipalmatis, an untidy, black and white bird with long legs and semipalmate toes. The face is featherless. The flight feathers are moulted progressively, avoiding a flightless period. These birds are vegetarian. The young are fed initially by the parents. Anseranas inhabit the floodplains of N. Australia.

Ant. About 15,000 species of social insect, related to bees and wasps. Ants live in large colonies and are easily recognizable by their slim 'waists', large heads and bent antennae. In a typical ant colony, the queen lays eggs which are tended by numerous wingless females called workers. In some species, there are large workers called soldiers; they have extra-large jaws and their task is to defend the nest or simply to crush hard foods. None of the workers lays eggs. Every year some of the eigs produce new queens and winged males, which mate during a 'wedding flight' from the nest. Afterwards the males die, but the queens either remain to increase the size of the existing nest or go off to found new colonies.

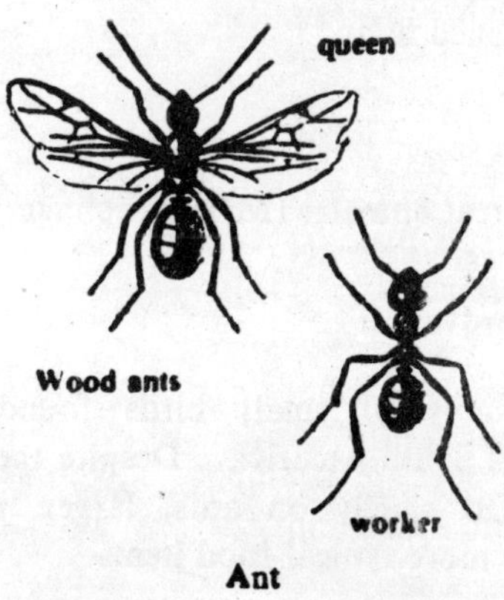

Ant

Many ants live on other insects but their diet is varied. Some 'milk' aphids for their honey-dew.

Antagonism

1. The interaction of two substances (*e.g.* drugs, hormones, or enzymes) having opposing effects in a system in such a way that the action of one partially or completely inhibits the effects of the other. For example, one group of anti-cancer drugs acts by antagonizing the effects of certain enzymes controlling the activities of the cancer cells.

2. An interaction between two muscles in which contraction of one (the antagonist) prevents that of the other (the agonist). The antagonist must relax to enable the agonist to contract and effect movement.

3. An interaction between two organisms (*e.g.* moulds or bacteria) in which the growth of one is inhibited by the other. Compare synergism.

Antagonist. A structure, process or reaction which is acting in opposition *e.g.* antagonist muscle (a muscle acting in opposite direction to the movement of another muscle).

Antarctic Cod. A common fish of antarctic waters named for its similarity to cod in taste. It has a large head and a small, flattened body.

Ant Bear

1. The great anteater (Myrmecophaga jubata) of South America.
2. The aardvark.

Antbird. A family of small birds found in the forests of Central and South America. Despite their name, antbirds do not feed chiefly on ants; larger insects, snails and worms are more typical food items.

Anteater. One of several mammals which feed principally upon ants, as the great anteater (Myrmecophaga jubata) of the order Edentata; the banded anteater (Myrmecobius fasciatus), a marsupial; the spiny anteater (Tachyglossus aculeatus), a monotreme; the scaly anteater or giant pangolin (Manis gigantea) of the order Pholidota.

Antebrachium. The forearm from elbow to wrist.

Antedonacea (feather-stars; subclass Articulata, order Comatulida). Superfamily of comatulids (feather stars) in which the variably shaped centrodorsal plate bears numerous small cirus sockets. The centrodorsal cavity is generally about one-third of the centrodorsal diameter. Five radial ossicles are present. The superfamily has existed from the Cretaceous to Recent.

Antelope. Any of a number of graceful, rapid-running ruminants of the family Bovidae. Term is applied to all gazelles and related forms (hartebeestes, duikers, waterbucks, sables, bushbuck, kudu) restricted entirely to the Old World.

Antelope. A group of hoofed mammals belonging to the same family as the cows, sheep and goats. They have horns which are unbranched and which do not fall off each year. Both males and females may have horns. Antelopes live in Africa and Asia, often forming huge herds. Gazelles, Oryxes, and Eland are all antelopes.

Antenna A long whiplike jointed mobile paired appendage which is present on the head of many arthropods. It is usually concerned with the senses of smell, touch, etc. In insects, millipedes, and centipeds they form the first pair of head appendages and are specialized and modified in many insects. In crustaceans they from the second pair of head appendages, the first pair (the antennules) having the sensory function, while the antennae are modified for swimming and for attachment.

Antennal Gland. A gland which opens on the antenna, as *a* that of certain insects; b the green gland of crustaceans.

Antennariidel anglerfish, frogfish; class Osteichthyes, subclass Actinopterygii, order Lophiiformes). Family of balloon-shaped, relatively short-bodied fish, with the naked and loose skin often possessing wart-like processes. The first of the three separate anterior dorsal spines is usually modified into a movable rod, the 'fishing pole' often provided with a fleshy lure used to attract prospective prey. The family comprises about 60 species of marine frogfish found in all tropical and subtropical seas.

Antennule. A small antenna, especially the more anterior and smaller of the two antennae found in certain crustaceans as the crayfish.

Antepenultimate. Second from the last segment, as in an insect antenna; third from the end.

Anterior

1. Designating the part of an animal that faces to the front, *i.e.* that leads when the animal is moving. In man and bipedal animals the anterior surface corresponds to the ventral surface.
2. Designating the side of a flower or axillary bud that faces away from the flower stalk or main stem, respectively. Compare posterior.

Anterior Cavity. In a vertebrate eye, the space containing aqueous humor located between the lens and cornea, subdivided into the anterior chamber, between cornea and iris, and posterior chamber, between iris and lens.

Anterior Choroid Plexus. The non-nervous tissue together with pia mater covering it forms villi-like processes which is extending into diocoel (cavity/ventricle of diencephalon) and is forming anterior choroid plexus; secretes cerebrospinal fluid.

Anterior Root. The ventral root of a spinal nerve composed of efferent fibers.

Anterior Vena Cava. A large vein which is collecting blood from anterior regions of the body; bilaterally symmetrical and opens into sinus venosus (in frogs, reptiles etc.) and sinuauricular opening in mammals.

Anthelate. Having elongated, flower-bearing branches, as in certain rushes.

Anthelmintic. A drug used to treat any worm infestation, especially one used to expel worms from the gastrointestinal tract.

Anther. The apical portion of a stamen, which produces the microspores or pollen grains. An anther normally comprises four pollen sacs (but only two in the Malvaceae) arranged in two groups or lobes joined by the connective tissue to the filament. If the anther is attached to the filament on the dorsal surface allowing it to pivot it is called a versatile anther. If it is attached dorsally but there is no movement it is termed a dorsifixed anther and if attached at the base it is a basifixed anther.

The wall of the anther lobes consists of an outer epidermis below which is the endothecium. This surrounds an inner nutritive layer or tapetum. Within the cavity (loculus) of the lobes pollen mother cells undergo meiosis to form tetrads of pollen grains. The individual pollen sacs are joined by zones of parenchyma tissue Compare ovule.

Anther Culture. The culture of excised anthers on sterile nutrient medium. If anthers are taken from a plant at a certain stage of development and cultured under appropriate conditions then embryoids and subsequently haploid plantlets may be induced to form from the pollen grains. In some species it is possible to culture isolated pollen grains. This precludes the possibility that any resulting plantlets might be derived from somatic tissue rather than a pollen grain. In culturing isolated pollen grains, anther tissue may have to be used as nurse tissue. If the ploidy level of plantlets derived from pollen grains is doubled using colchicine then completely homozygous diploid plants can be obtained. Compare ovule culture.

Antheridial Cell. The cell from which the antheridium develops. In seed plants, it is the generative cell in the pollen grain, which divides to provide the two sperm cells in the pollen tube. Compare prothallial cell.

Antheridiophore. An upright structure consisting of a stalk and cap that bears the antheridia in certain liverworts of the Marchantials. The antheridia are borne in pits on the upper surface of the cap. Compare archegoniophore.

Antheridium. The male sex organ of algae, fungi, bryophytes, and pteridophytes. It produces the male gametes (antherozoids). It may consist of a single cell or it may have a wall that is made up of one or several layers forming a sterile jacket around the developing gametes. Compare archegonium.

Antheriferous. Bearing anthers.

Antherocyte (spermatocyte). A cell that differentiates into an antherozoid without further cell divison. The antherozoid is usually released from the antherocyte after it has been discharged from the antheridium.

Antherozoid (spermatozoid). A motile male gamete produced by lower plants and some gymnosperms, which moves by means of flagella. In lower plants antherozoids are released from an antheridium but in the gymnosperms they are formed in the pollen tube prior to fertilization. Most antherozoids consist of an elongated nucleus contained within a ribbon-like cell. This form enables them to penetrate the narrow neck of an archegonium. The number of flagella borne by an antherozoid may be used as a diagnostic character.

Anthesis. The period from flower opening to fruit set.

Anthocephalus (family Rubiaceae). Genus of fastgrowing, medium-sized, pioneer trees of lowland rain and seasonal forests, characteristically of moist alluvial soils. They yield useful timber, and have been planted widely, but mainly unsuccessfully, throughout the humid tropics. There are two species, native to Nepal, and to the region from Yunnan to New Guinea.

Anthocarpous. Pertaining to aggregated fruits with accessory structures called pseudocarps, as in the strawberry.

Anthocaulus. In solitary corals, the stalk after separation of the anthocyathus.

Anthocerotae (hornworts, horned liverworts). A class of the Bryophyta containing a single order, Anthocerotales, and some five genera. Members of this order are distinguished from those of liverwort orders, with which they are sometimes classified, by their long-lived axial sporophytes. These contain photosynthetic tissue and stomata and occasionally persist after the death of the thallose gametophyte. Most species are also unusual in having a single pyrenoid-containing chloroplast in each cell, a feature reminiscent of the algae.

Anthocharis Cardamines (orange-tip butterfly; order Lepidoptera, family Pieridae). Butterfly in which the forewings of the males are white, tipped with black, and with a large, orange, apical patch. The orange is absent from females. Larvae feed on Cruciferae, and are well camouflaged against the food plant. The species is represented in Europe, including Britain, and in temperate Asia.

Anthochlor Pigments. A group of yellow flavonoid flower pigments containing the chalcones and aurones. They turn red on exposure to ammonia. This reaction distinguishes them from the yellow carotenoid pigments, which do not react in this way.

Anthocodium. The oral end of an alcyonarian coral.

Anthocyanescence. The development of red pigments as a symptom of disease as, for example, seen in peach leaf curl caused by the fungus Taphrina deformans.

Anthocyanin. One of a group of flavonoid pigments. Anthocyanins occur in the cell vacuoles of various plant organs and are responsible for many of the blue, red, and purple colours in plants (particularly in flowers).

Anthocyathea (Irregulares; phylum Archaeocyatha). Class of solitary, rarely colonial invertebrate animals found in Lower, Middle, and Upper Cambrian rocks. The conical cup is from cylindrical to discoid in outer form, often with an irregular outline, having one, or more usually two, porous walls. The intervallum contains rods and bars or septa, always with dissepiments and commonly with tabulae.

Anthocyathus. In the young of solitary corals, the disklike terminal portion which separates from the stalk or anthocaulus.

Anthodium. A composite flower; head; capitulum.

Anthogenesis. The production of both males and females by parthenogenic females, as in aphids.

Anthophilous. Attracted to flowers; feeding upon flowers.

Anthomyiidae (class Insecta, order Diptera, suborder Cyclorrapha). Moderately-sized family of small to tiny flies with slender bodies, which may be recognized by the position of the posterior crossvein in the middle of the wing, the pubescent third antennal segment, and costal cilia. Vein Cu_1 and Cu_{1a} reaches the wing margin, but may be faint at the extremes. The larvae are phytophagous, feeding in plant shoots or inside galls. Some genera, *e.g.*, Erioischia and Hylemyia, are economically important as pests of cabbage and wheat respectively. More than 1000 species have been described.

Anthophoridae (order Hymenoptera, suborder Apocrita). Large, diverse, cosmopolitan family of mainly solitary bees, with a few species which show some degree of sociality. Most species are longtongued and have a rapid, darting fight. A pygidial plate is present in the females of almost all species, and in most males. The clypeus is usually protuberant, and the anterior coxa is only slightly broader than long. The pollen scopa consists of hairs and is restricted to the hind tibiae and basitarsi. The family includes some of the largest bees, *e.g.*, the neotropical genera Centris and Epicharis, and the pantropical carpenter bees of the genus Xylocopa. Other genera include Anthophora, Amegilla, Eucera, Tetralonia, Synhalonia, Exomalopsis, Tetrapedia, and Paratetrapedia. With the exception of the carpenter bees (Xylocopa, Ceratina, and related, genera), which bore into solid wood or plant pith, all anthophorides are ground nesters. They line their brood cells with a waterproofing secretion of the Dufour's gland. In at least one species of Anthophora this comprises a mixture of triglycerides, which is also mixed with the larval food. Some species of Centris and Epicharis collect plant oils, as well as pollen and nectar, as larval food. One subfamily, the Nomadinae, consists entirely of parastic or cuckoo bees, the females of which have lost the pollen scopa and lay their eggs in the nests of other bee species. The Nomadinae include the genera Nomada, Ammobates, Epeolus, Passites, and Neolarra.

Anthophyte. Any member of the anthophyta, a division of the plant kingdom which includes all flowering plants, vascular plants in which double fertilization occurs resulting in development of a fruit containing seeds. It comprises two classes, the Monocotyledonae and Dicotyledonae.

Anthotaxy. The arrangement of flowers in a cluster.

Anthoxanthin. Any of a class of yellow or cream glycoside plant pigments normally consisting of a glucose molecule attached to a flavone or xanthone molecule.

Anthozoan. Any coelenterate of the class Anthozoa which includes the sea anemones and corals.

Anthracnose. A fungal plant disease in which the characteristic symptoms are limited lesions, necrosis, and hypoplasia. Anthracnose diseases are generally caused by one of the Melanconiales (*e.g.*, Colletotrichum lindemuthianum, bean anthracnose; C. coffeanum, coffee anthrcanose or coffee-berry disease; and Gloeosporium limetticola, lime anthracnose). The fungi causing anthracnose diseases produce numerous spores that are spread by rain. High humidity is required for infection and the diseases are most destructive when there is some water-soaking of tissues. Anthracnose diseases also produce symptoms, including scab, leaf-spots, and blight, that are not exclusive to the group.

Anthracotheriidae. Large, primitive members of the Artiodactyla, especially common in the middle Tertiary of the Old World. They were pig-like in general appearance. From the nature of the sediments in which they are fossilized, it seems that they were amphibious. This supports the view based on anatomical evidence that they were the ancestors of the hippopotamuses. They arose in the Eocene and became extinct in the Pleistocene.

Anthracothorax (mangoes; family Trochilidae). Genus of large hummingbirds with green backs and blackish under-parts in males. The bills are slightly decurved. They are nectar feeders, but will take insects. There are seven species, f[illegible] C. and S. America, of which three species ar[illegible] endemic to W. Indian islands.

Anthrax. Disease of animals, including humans, caused by the bacterium Bacillus anthracis. The most common form of the disease is the anthrax boil, a localized boil or pustule resulting from infection of a wound by B. anthracis. There is also a systemic form in which septicaemia occurs; this form is frequently fatal. Anthrax is uncommon in humans, occurring mostly in people who handle animals or animal products.

Anthreptes (sunbirds; family Nectariniidae). Genus of small sunbirds, which are less specialized than most members of the family, and have relatively shorter and straighter bills. They feed on nectar, small insects, and fruit. A. pallidigaster (Amani sunbird) is endangered, and confined to two forest areas of Kenya and Tanzania. There are 17 species, found in Africa, SE Asia, and Indonesia.

Anthribidae (class Insecta, order Coleoptera). Family of robust beetles, 2–15 mm long, with a short, flattened rostrum. The antennae are short and clubbed, or filiform and very elongate, but without scape. Larvae are curved and fleshy; with legs and antennae absent or reduced. They feed inside fungi, dead wood, and plant galls. Doticus pestilans develops inside dried apples. Araecerus fasciculatus is a pest of coffee beans, cocoa beans, and spices, particularly nutmeg. Brachytarsus larvae feed parasitically inside scale insects. There are 2400 species, mostly tropical.

Anthriscus (family Umbelliferae). Genus of herbaceous, fern-like plants, with two or three pinnate leaves, and compound umbels. Usually there are no bracts, but several bracteoles. The flowers are small, and white. The fruits are ovoid to oblong, smooth or spiny, and hardly ridged. A. cerefolium (chervil) is cultivated as a vegetable. There are some 12 species, found in temperate Eurasia, and in N. Africa.

Anthropomorphic. Pertaining to human characteristics of other organisms.

Anthropoidea (Simiiformes; cohort Unguiculata, order Primates). Suborder comprising the monkeys, apes and humans. Monkeys and apes have a common ancestor and diverged in the Oligocene. The Dryopithecines of the succeeding Miocene were undoubted apes.

Anthropogenic. Caused by or resulting from the acts of man.

[1]Anthropoid. Manlike.

Anthropoid. Any primate of the suborder Anthropoidea which includes the monkeys, apes, and man.

Anthropoid Ape. One of the large, tailless apes of the family Pongidae, as the gibbon, chimpanzee, orangutan, or gorilla.

Anthropology. The science which deals with man.

Anthropomorphic. Attributing human attributes to nonhuman beings or objects.

Antiaris (family Moraceae). Genus of trees producing poisonous latex which is much sought as a constituent for arrow poisons. A. toxicaria (ipoh or upas tree) gained a false reputation in Europe for being excessively poisonous. There are four species, occurring in tropical Africa, Madagascar, and Indo-Malaysia.

Antibiosis. An association between two organisms in which one produces a substance or a condition which is harmful to the other.

Antibiotics. Substances obtained from microorganisms especially moulds, that destroy or inhibit the growth of other microorganisms, particularly disease-producing bacteria and fungi. Common antibiotics include penicillin, streptomycin, and tetracyclines. They are used to treat various infections but tend to weaken the body's natural defence mechanisms and can cause allergies. Overuse of antibiotics can lead to the development of resistant strains of microorganisms.

Antibody. A protein (immunoglobulin) produced by certain white blood cells (lymphocytes) in response to entry into the body of a foreign substance (antigen) in order to render it harmless. An antibody—antigen reaction is highly specific. Antibody production is one aspect of the immune response and is stimulated by such antigens as invading bacteria, foreign red blood cells (ABO system), inhaled pollen grains or dust, and foreign tissue grafts.

Antibrachium. Anterior part of fore limb that is joined anteriorly with upper arm and posteriorly with hand, supported with radius and ulna bones. Also known as forearm.

Anticlinal. At right angles to the surface. The anticlinal wall of a cell is thus arranged perpendicular to the surface of the plant body. An anticlinal division results in the formation of anticlinal walls between daughter cells. Such a division enables a tissue to increase its circumference, thus keeping pace with any increase in girth of the organ. In cylindrical organs, such as stems and roots, the term radial may be used in place of anticlinal, especially when describing cell walls. Compare periclinal.

Anticoagulant. A substance that does not allow the formation of blood clots. Heparin is a natural anticoagulant, which is extracted to great conditions like thrombosis and embolism. Synthetic anticoagulants are warfarin and dicoumarol.

Anticodon. A sequence of three nucleotides on transfer RNA that is complementary to a sequence (the codon) on messenger RNA, to which it temporarily binds during protein synthesis. A given molecule of transfer RNA will possess a specific anticodon that only complexes with one particular amino acid. It is this absolute correspondence between an amino acid, an anticodon, and a codon, that enables the type and sequence of amino acids in a protein to be determined precisely.

Antidesma (family Stilaginaceae). Genus of shrubs or small trees producing flowers on spikes, which are minute, unisexual, and have a prominent disc. The fruit is a small, juicy drupe, usually flattened. The genus is often placed in Euphorbiaceae, but has no close relatives there, and perhaps is intermediate to Icacinaceae. There are 170 species, occurring in the Old World subtropics and tropics, especially in Asia.

Antidiuretic. Opposing or preventing the formation and excretion of urine.

Antidiuretic Hormone. Vasopressin—also called ADH.

Antidiuretic Hormone. A substance which is secreted by the nerve endings passing from the hypothalamus into the posterior pituitary lobe in a vertebrate. It is able to control the absorption or resorption of water by the kidney tubules. Insufficiency of this hormone brings about an increase in water intake resulting in diabetes inspidus.

Antidromic. Moving in a direction opposite to normal, as nerve impulses travelling peripherally in afferent fibres.

Antidromous. *bot.* Spiraling in a reverse direction.

Antienzyme. A substance inhibiting the normal action of an enzyme without actually destroying it. Such substances are given out by the epithelium of the stomach, and prevent its being digeste by its own enzymes.

Antigen. Any substance that the body regards as foreign and that therefore elicits an immune response. Antigens may be formed in, or introduced into, the body. They are usually proteins. Transplantation (histocompatibility) antigens are associated with the tissues and are involved in tissue or organ grafts; an example is the HL-A system. A graft will be rejected if the recipient's body regards such antigens on the donor's tissues as foreign.

Antihistamine. Any drug that inhibits the effects of histamine in the body and is therefore used to relieve and prevent the symptoms associated with allergic reactions, such as hay fever. Since one of the side-effects produced by antihistamines is sleepiness, some are used to prevent motion sickness and induce sleep.

Antilocapridae (order Artiodactyla, infar-order Pecora, superfamily Bovoidea). Family of 'antelopegoats', which appeared first in the middle Miocene. In general appea rance, the earliest representatives probably resembled roe deer. The horns have two branches in bucks, and are usually simple in does but sometimes have a small prong. The horns have a hard bony core, and a soft, keratinous covering which is shed. The feet posscess two digits. Antilocaprids graze or browse in herds. There is one surviving species, in one genus (Antilocapra, the pronghorn), found in western N. America

Antilymphocyte Serum (ALS). A serum used to suppress the immune reaction in patients who are receiving tissue or organ transplants. It is prepared by injecting human lymphocytes into a horse, which then produces antibodies against them. ALS is then obtained from the horse and the antibodies are extracted and purified. Now this preparation is injected into the transplant patient when it destroys his lymphocytes, which would otherwise have produced antibodies causing rejection of the transplant.

Antimere. One of a pair of corresponding parts in a bilaterally symmetrical animal.

Antimetabolite. A substance which interferes with normal cellular metabolism, especially one which inhibits or blocks a specific step in a metabolic pathway. Antimetabolites include sulfonamides, antibiotics, antihistamines, and various herbicides and insecticides.

Antimycin. An antibiotic which is isolated from Streptomyces strain. It blocks electron transport system by inhibiting transport of electrons from cytochrome b to cytochrome-c.

Antipatharia (black corals; class Anthozoa). Order of colonial, branching corals. Polyps are small, with few tentacles (six simple, or eight branched) and few unpaired mesenteries (6, 10, or 12). The skeleton is a branched, thorny, chitinous, axial structure, developed from the ectoderm, extending throughout the colony, and surrounded by a fleshy coenosarc containing the polyps. Black corals are non-reef-dwellers in depths of 2-35 000 m. About 20 extant genera are known, one of which is also known from the Miocene.

Antipodal. Located on the opposite side.

Antipodal Cells. The haploid cells, usually three in number, found in the embryo sac at the opposite end to the micropyle. They are derived by mitotic divisions of the megaspore and have no distinct cell wall. They take no part in the fertilization process and their function is unknown. At fertilization they may disintegrate or multiply and enlarge.

Antipodal Cells. The three haploid cells which are found in the embryo sac of seed-bearing plants that migrate to the chalazal end of the sac opposite the micropyle. These nuclei arise due to the three meiotic divisions that produce the egg cell, synergid cells, and polar nuclei, but they themselves are not taking part in the fertilization process. They are eventually absorbed by the developing embryo, and their function is not known with certainty.

Antirrhinum (family Scrophulariaceae) Genus of plants in which the lower leaves are opposite, the upper leaves alternate, and all are simple. The flowers are borne in racemes, and have strongly two-lipped corollas, the lower three-lobed lip having a projecting bulge or 'palate' which closes the mouth of the flower; the tube is unspumed, but with a swelling below. A. majus (snapdragon) is cultivated in many coloured forms and has been naturalized beyond its S. European natural range. There are some 40 species of herbs, most Mediterranean, but also occurring in western N. America.

Antisepsis. Prevention of decay, putrefaction, or sepsis by exclusion of microorganisms or inhibition of their growth and multiplication.

Antiseptic. Any substance that kills or inhabits the growth of disease-causing microorganisms but is essentially nontoxic to cells of the body. Common antiseptics include hydrogen peroxide, the detergent cetrimide, and ethanol. They are used to treat minor wounds. Compare disinfectant.

Antiserum. An immune serum prepared from the blood of man or an animal which has been immunized against bacteria or other antigenic agents and which contains antibodies specific for these agents.

Antithetic. Heteromorphic.

Antithrombin. A substance in blood plasma which neutralizes thrombin, thus preventing coagulation.

Antitoxin. A substance which neutralizes a toxin, especially an antibody which neutralizes or inactivates the toxin that serves as its antigen.

Antitranspirant. A compound that reduces transpiration, either by closing the stomata or by depositing a film over the stomata. Antitranspirants are sprayed on some crops and ornamentals, but their use also limits photosynthesis and affects various other metabolic processes.

Antler. Bony growth from the skull of deer, usually branched and initially covered in fine skin called velvet. Antlers grow in pairs and are shed each year. Each year's antlers are larger than those of the year before, until the animal reaches maturity. After this they remain the same size for a few years, and then may actually decline. Except in the case of reindeer, antlers are confined to the males.

Ant Iron An insect resembling a dragonfly but more closely related to the lacewing fly. The name 'ant lion' refers to the young insects. They live where the soil is sandy and some species make little pits for themselves. They lie more or less buried at the bottom of the pit and wait for an ant or some other insect to tumble down the loose sandy sides. They then pounce on it and eat it. Other species merely lie buried in the sand with just their great jaws exposed, ready to grab any passing insect. Ant lions live in most of the warmer parts of the world.

Antrorse. Directed upward and forward. Compare retrorse.

Antrum

1. A hollow space or cavity, especially one in bone.
2. The cavity cf an ovarian follicle.

Antrum Folliculi. A space/cavity which is formed in between cumulus cophorous (cell surrounding immature ovum) and stratum grauulosum (outer cells) of Graffian follicic; filled with a liquid which is known as liquor folliculi.

Antrum of Highmore. The maxillary sinus.

Anuclear. Lacking a nucleus, as an erythrocyte.

Anura (Salientia; anurans, tailless amphibians, frogs and toads; class Amphibia). Order of amphibians in which the head joins directly into trunk. The vertebral column is short, and there is no tail. The forelimbs are stout; the hind limbs long and webbed, and used for jumping and swimming. The eyes are large, and high on the head. There are superficial ear membranes posterior to the eyes. All anurans lay eggs, have external fertilization, and develop from tailed tadpoles. There is no neotony. Sexual dimorphism is common, the males often being smaller and having vocal pouches. There are 2600 species, widespread in tropical and temperate zones.

Anuran. A salientian; a frog or toad.

Anuria. Inability to form or excrete urine.

Anus. The terminal opening of the alimentary canal in most animals, through which indigestible material (facces) is expelled.

Anvil. The incus; the middle ear bone.

Aorta. The major blood vessel in higher vertebrates through which oxygenated blood leaves the heart from the left ventricle. The aorta branches to form many smaller arteries, which in turn branch many times to supply oxygen and essential nutrients to all living cells in the body.

Aorfic Arch

1. In vertebrate embryos, one of a series of paired blood vessels which encircle the pharynx connecting the ventral and dorsal aortae.
2. In aquatic vertebrates, one of a series of paired afferent and efferent branchial arteries.

Aortic Body. Refers to a nodule on the aortic arch which is having chemoreceptors that sense the composition of blood; it monitors changes in O_2 and CO_2 in arterial blood *e.g.* carotid body of mammals and birds, carotid labyrinth of amphibia. The chemoreceptors have been innervated by branches of IX (glossopharyngeal) and X (vagus) cranial nerves.

Aortic Ring. A ring of dense fibrous connective tissue at the orifice of the aorta in the heart of certain mammals, sometimes containing a heterotopic bone, the os cordis, as in cattle.

Apara The three-banded armadillo (Tolypeutes tricinctus) of South America.

Apatite. A complex mineral form of calcium phosphate, $Ca_5(PO_4)_3(OH,F,Cl)$, that is the main constitutent of the enamel of teeth.

Ape

1. In a broad sense, any monkey.
2. One of the large, tailless, semierect primatcs of the family Pongidae.

Apeture. An opening or orifice.

Aperturate. Describing a pollen grain having one or more apertures (areas where the exine is either thinner or absent). If the apertures are colpi the pollen is termed colpate and if they are pores, porate.

Aphagia. Inability to swallow.

Aphaniptera (Siphonaptera). The order of insects that includes, the fleas, which are all ectoparasites of mammals and birds, *e,g.*, Pulex irritans (human flea). Some are able to transmit serious diseases; for example Xenopsylla (the rat flea) carries bubonic plague. Usually each species is limited to one host.

Aphasmidia. A subclass of Nematoda incorporating the whipworm, kidney worm and trichina. They are characterized by the absence of phasmids or percaudal sensorp organs, but they are having amphids or anterior chemoreceptors.

Apical. The end or ontermost part as of a conical structure.

Aphid. 'Greenfly' or 'blackfly' are two of the commonest of the many types of aphid. Also called plant lice, aphids are sap-sucking bugs and they do an immense amount of damage to crops by sucking sap and by spreading germs that cause plant disease. Most aphid species have both winged and wingless forms.

Aphididae (aphids, greenfly, blackfly; order Hemiptera, suborder Homoptera). Family of homopterans which feed on plant sap. They are soft bodied insects, with the rostrum arising between the front pair of legs. Typically there is only one generation a year that reproduces sexually, laying overwintering eggs, usually on woody plants; and several generations of asexually and viviparously reproducing females, either on the woody host or on herbs. Many species are tended by ants which feed on the honeydew (sap that has passed through the bodies of the aphids). Some aphids are serious pests. There are about 4000 species, occurring mainly in the temperate northern hemisphere.

Aphodus. A narrow canal connecting a flagellated chamber with an excurrent canal in certain leuconoid sponges.

Aphonia. Loss of speech; voicelessness.

Aphotic. Without light, as in the aphotic zone of the ocean, the region below 800 meters which receives little or no light.

Aphyllous. Without leaves.

Aphytal. Lacking plants.

Apian. Of or pertaining to bees.

Apiary. A place where bees are kept, especially a collection of hives maintained for honey production.

Apical. At, near, or belonging to an apex, tip, or summit.

Apical Dominance. The inhibition of the development of some or all of the lateral buds by the terminal (apical) bud of a shoot. Removal of the terminal bud releases some of the lateral buds from inhibition. This implies that a substance produced at the apex, most probably auxin, is responsible for the inhibition, though its method of action is nuclear. Apical dominance is not so marked when nutrients are plentiful, suggesting that available nutrients are first delivered to the terminal bud and only those in excess of requirements reach the lateral buds. Cytokinins have also been shown to promote the growth of lateral buds.

Apical Meristem. A region at the tip of each shoot and root of a plant in which cell divisions are continually occurring to produce new stem and root tissue, respectively. The new tissues produced are known collectively as the primary tissues of the plant.

Apical Organ. A sensory structure at the apex of a trochophore which bears a tuft of cilia.

Apical Placentation (pendulous placentation, suspended placentation). A form of placentation, found in ovaries containing only one ovule, in which the placenta develops at the top of the ovary.

Apiculate. With a short, abrupt point.

Apiculture. Bee culture.

Apidae (bumblebees, honeybees, orchid bees, stingless bees; order Hymenoptera, suborder Apocrita). Family of bees, in which the pollen-transporting apparatus is on the outer face of the hind tibia and comprises a corbiculum (a polished, slightly concave area bounded by stiff bristles). The hind tibial spurs are absent, except in the genus Bombus. Genera include the neotropical orchid bees Eufriesia, Eugloss, and Eulaema, and the primitively eusocial Bombus (bumblebees). The stingless bees of the genera Melipona and Trigona; and the four species of honeybees—Apis cerana, A. dorsata, A. indica, and A. mellifera—have achieved the highest levels of insect sociality other than those found among the termites and ants, with a morphologically distinct worker caste.

Apionidae (weevils; class Insecta, order Coleoptera) Insects very closely related to Curculionidae, but distinguished by having non-elbowed, clubbed antennae. They have a highly convex, punctured elytra, 1.5-20 mm long; and the head is produced into a long, curved or stout rostrum. The larvae are legless and grub-like, and feed inside seed pods, stems or plant roots. Apion larvae are often pests of legumes : females bite a hole in the pod, lay eggs on developing seeds, and plug the hole. Cyclas formicarius, the sweet-potato weevil, is a crop pest. There are 1060 species.

Aplacophora (Solenogastres; phylum Mollusca, class Amphineura) Subclass of benthic, worm-like molluscs, which lack a dorsal covering of valves. Calcareous spicules may occur in the mantle that covers the body. The anatomy and nervous system are very similar to those of the Polyplacophora. The ventral foot is represented as a groove in the order Neomeniida, but is absent in the second order, Chaetodermatida. They are entirely marine, and have a worldwide distribution. This group does not occur in the fossil record.

Aplanospore. A non-motile spore. It is characteristic of the pin moulds and green algae. It is an asexual spore which is formed in a sporangium and is usually thick walled.

Aplanetic. Describing organisms in which there is no motile stage.

Aplanospore. A nonmotile spore as produced, *e.g.*, certain algae in the chlorophyta and chrysophyta.

Aploactinidae (Haploactinidae; velvetfish; class Osteichthyes, subclass Actinopterygii, order Scorpaeniformes). Family of marine, coastal fish of the W. Pacific. The sharply tapering body shows a long dorsal fin originating on the head, with the pelvic fins reduced to a weak spine and two or three rays. The scaleless skin bears papillae, giving it a velvety appearance. They are relatively common in weedy areas of Australia. Aploactisoma milesii may reach a length of 23 cm.

Aplochitonidae (Haplochitonidae; southern smelts; class Osteichthyes, subclass Actinopterygii, order Salmoniformes). Family of freshwater and brackish-water fish with an elongate body, and a single dorsal as well as an adipose dorsal fin. The Australian grayling, Prototroctes maraena, and the Derwent whitebait, Lovettia sealii, ascend the rivers in order to spawn.

Apne. Refers to the suspension or absence of breathing, brief periods of respiratory arrest. May be generated by injecting varatridine and nicotine into coronary and pulmonary circulation thereby producing bradycardia and hypotension and due to stimulation of chemoreceptors.

Apneumatic. Not filled with air.

Apneusis. If all cranial nerves are cut and brain stem gets transected above pons, regular breathing continues. However, an additional transection in the inferior portion of pons causes inspiratory neuron discharge continuously resulting in a sustained contraction of inspiratory muscles and such arrest of respiration during inspiration is termed as apneusis.

Apneustic Centre. When vagi are cut and animal develops a peculiar type of breathing in which prolonged respiratory cycle is spent in inspiration and after a brief expiration there occurs a sustained inspiration. Such a breathing occurs due to apneustic centres that are existing in pontine area of vagus. The centre discharges tonically upon medullary respiratory centre thereby promoting more sustained respiratory activity.

Aplodontidae (mountain beaver; order Rodentia, suborder Sciuromorpha). Family of burrowing rodents. The body is thickset and heavy, the limbs and tail short, the claws large. The eyes and ears are small. The skull is flat, wide posteriorly, and the palate broad. The animal is found only in the N. American Rockies. There is one genus (Aplodontia), and one species.

Apnenstic. Lacking spiracles, or with nonfunctional spiracles, as in aquatic insects.

Apocarpous. Describing a gynoecium in which the carpels are free, as in buttercups (Ranunculus). This type of arrangement is thought to be more primitive than a syncarpous gynoecium.

Apetalous (of a flower). Without petals.

Apatetic Coloration. Coloration which helps to protect an animal by misleading its enemies. In some cases the outline of the animal has been broken up (Disruptive Coloration); in others the makings reveal the presence of an eye or other organ where in fact there is none.

Aperture. Refers to orfice or opening.

Apex

(a) The summit or top; the vertex or summit;

(b) The tip of a conelike structure; the end opposite the base.

Apocarpy. The condition in which the female reproductive organs (carpels) of a flower are not joined to each other. It occurs, for example, in the buttercup. Compare syncarpy.

Apocrine. Refers to a gland in which the products are collected at the outer part of cells and squeezed out with a little cytoplasm thereby a part of cells gets damaged. For example, the sweat glands of mammals.

Apocrine Gland. A gland in which a portion of the cytoplasm of the secretory cells is lost and becomes a part of the secretion, as the mammary gland.

Apocrine Secretion. Secretion

Apocrita (ants, bees, wasps; subclass Pterygota, order Hymenoptera). The larger of the two suborders of Hymenoptera, whose members have a constriction between the thorax and abdomen, occurring between the narrowed first abdominal segment (which is attached to the thorax) and the rest of the abdomen. Wing venation in some minute forms is greatly reduced, and the ovipositor can be of two types; a non-stinging type normally kept outside the abdomen, and a stinging type which is retractable. Apocrita is a very large suborder, comprising 11 superfamilies, with about 1 05 000 species, many of which are of benefit to humans. Species whose larval stages are herbivorous include bees, gall wasps, and some Chalcidoidea, but the majority of Apocrita are parasites of other insects and spiders.

Apocynaceae. Family mostly of trees, climbers or shrubs. The leaves are mostly opposite, or whorled. The flowers are cymose, regular, and bisexual; they have a five-lobed calyx and corolla tubes, five stamens, the ovary superior or half inferior, and two carpels, which are free or fused. Fruits are paired, fleshy or woody, and dehiscent. There are about 180 genera, with about 1500 species, whose distribution is tropical with a few temperate outliers.

Apodal

1. Lacking feet or legs; apodous.

2. Lacking pelvic fins, as eels.

Apodeme. An inward projecting portion of the exoskeleton of an arthropod serving primarily for the attachment of muscles.

Apodidae (swifts; order Apodiformes). Family of medium to small birds with small, short bills and a wide gape. They have long, curved narrow wings; a short tail, often spine-tipped; and short, weak legs with reversible hallux. They are highly aerial, fast-flying, and insectivorous. Their nests are built on trees, buildings, rocks, and in caves, often using saliva. There are 12 or more genera, containing about 80 species. Nearly all are tropical, northern, breeding birds migrating to the tropics in winter, but they are found worldwide.

Apodous

1. Lacking feet; apodal.
2. Without thoracic or abdominal locomotor appendages, with reference to insect larvae.

Apoenzyme. The catalytically inactive protein portion of an enzyme that remains when the prosthetic group or cofactor has been removed. Examples of enzymes needing both apoenyme and cofactor for catalytic activity include; alcohol dehydrogenase, which requires zinc ions; kinases, which require magnesium or manganese ions; and the cytochromes, which require ferrous, Fe (II), or ferric, Fe(III), ions.

Apogamy. In pteridophytes, the term refers to the development of the sporophyte directly from a cell of the gametophyte, so fusion of gametes is bypassed. It frequently takes place in gametophytes that have been produced aposporously and are thus diploid. The term is also used to describe the development of an unfertilized female gamete into the sporophyte, a phenomenon called parthenogenesis.

Apogonidae (cardinal fish; class Osteichthyes, subclass Actinopterygii, order Perciformes). Large family of mainly marine fish not exceeding 20 cm in length. They exhibit two separate dorsal fins, and have relatively large eyes and mouth. Most cardinal fish display the unusual habit of incubating the fertilized eggs in the mouth cavity until they hatch. In some species the males are the mouth-brooders, in others the female performs this duty. There are about 170 species. They are shallow-water fish of all temperate and tropical seas.

Apollo. A medium-sized butterfly found on mountains in

Apollo

Apollo

mainland Europe. There are several races, mostls with white, yellow or grey wings and dark markings. The large spots on the hind wing are usually red or orange.

Apollo Butterfly. A group of mountain-dwelling butterflies related to the swallowtails. Found in Europe. Asia and North America, they are mostly white with black or red markings.

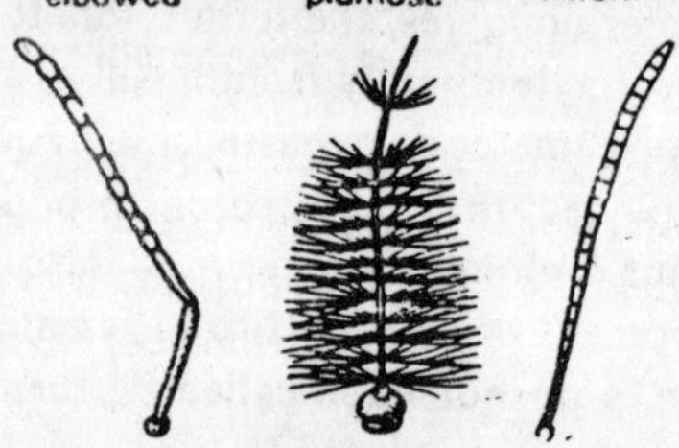

Various types of antennae

Apolysis. The successive breaking off of gravid proglottids in tapeworms.

Apomixis (agamospermy). A reproductive process in plants that superficially resembles normal sexual reproduction but in which there is no fusion of gametes. In apomictic flowering plants there is no fertilization by pollen, and the embryos develop simply by division of a diploid cell of the ovule.

Apomorphy. In discussions of phylogeny or cladistics, a derived or advanced character state. Apomorphies shared by different taxa are termed synapomorphies (compare autapomorphy). Only synapomorphies whose origin can. be traced back to a recent common ancestor (*i.e.* homologous character states) are of use in constructing phylogenies or cladograms. Analogous 'synapomorphies' may arise by convergent evolution and if identified as such are ignored. Compare plesiomorphy.

Aponeurosis. A flat, sheetlike tendon which attaches muscle to bone or other structures.

Apopetalous. With separate and distinct petals; polypetalous.

Apophyses. Refer to various kinds of projections that are arising from a vertebra *i.e.* zygapophyses, tranverse processes etc.

Apophysis. The slightly swollen region between the seta and the capsule in a moss sporophyte. Its cells are rich in chloroplasts, there are numerous intercellular spaces, and the epidermis contains stomata, similar in structure to those of vascular plants. The apophysis is thus an active photosynthetic region and helps to nourish the developing sporogonium.

Apoplast. Refers to the system of cell walls which are extending through a plant body. Along this water having mineral salts, etc. can move passively. It is an important pathway for moving these substances outside the xylem, for example across the root cortex.

Apopyle. In sponges, the opening of a radial canal into the central cavity or spongocoel; in elasmobranchs, the anterior opening of the clasper tube. Compare hypopyle.

Aporogamy. Fertilization in plants in which the pollen tube does not enter through micropyle. Compare porogamy.

Aposematic. Warning or threatening, with reference to coloration.

Aposematic Coloration. Warning coloration.

Aposepalous. With separate and distinct sepals; polysepalous.

Apospory. The development of the gametophyte from the sporophyte without meiosis and spore production, so that the gametophyte has the same number of chromosomes as the sporophyte. If fertile gametes are formed by such a gametophyte then a sporophyte with twice the original number of chromosomes is produced. A polyploid series may be built up in this way. Apospory can be induced artificially in certain ferns, *e.g.* the lady fern

(Athyrum filix-femina), by pinning segments of frond to damp sand. Prothalli then arise from buds on the frond.

Apospory may also be used to describe the condition in angiosperms in which a diploid embryo forms from a cell of the nucellus or chalaza and megaspore formation is bypassed.

Apostasiaceae. Family of monocotyledonous herbs with plicate leaves. The flowers are racemose, with six free perianths, 2-3 stamens, the filaments and style, united, the ovary inferior and trilocular, and numerous axile ovules. The fruit is a capsule. The family is primitive, and closely related to Orchidaceae. There are two genera, with about 20 species, found from India to tropical Australia.

Apostatic Selection. Selection favouring diversity in appearance and genetic variability among the individuals of a species population. Mimicry, for example, is a polymorphism in which individuals of a species that is palarable to predators mimic other species that are unpalatable. Because it generates population diversity, a typicallity is a positive advantage to individuals and to the species overall : the fewer mimics there are in relation to the numbers of their genuinely unpalatable models, the more frequently the predators will repeat the unpleasant experience of eating an unpalatable individual. This reinforces the association of appearance with unpalatability. Predators will tend to avoid apparently unpalatable individuals. This will favour reduced predation of the mimic species. Were all individuals to mimic the same model, however, this advantage would be lost, as the proportion of mimics to models would rise. It would then be less common for predators to encounter unpalatable prey and the reinforcement would be reduced. Thus by mimicking a variety of models, the population of the mimic species may be increased.

Apothecium. The open disk or cup-shaped fruiting body or ascocarp of certain ascomycetes, as the cup fungi.

Apparatus. A group of structures or parts which work together in performing a certain function, as the Golgi apparatus.

Appeasement. Behaviour that inhibits aggression from another animal of the same species, frequently taking the form of a special posture or display emphasizing the weakness of the performer. Threatening structures (*e.g.* antlers) and makings are covered or turned away, and vulnerable parts of the body may be exposed. Appeasement is seen in courtship, in greeting ceremonies, and often (from the loser) after a fight.

Appendage. A structure attached to or appended to a larger structure; a projection of the body or of an organ; a limb.

Appendicular. Consisting of, pertaining to, or possessing appendages.

Appendicular Skeleton. Constituting collectively the skeletal elements of girdles and limbs; includes pectoral and pelvic girdles.

Appendiculate. Provided with or bearing an appendage appendages.

Appendix (vermiform appendix). An outgrowth of the caecum in the alimentary canal. In humans it is a vestigial organ containing lymphatic tissue and serves no function in normal digestive processes. Appendicitis is caused by inflammation of the appendix.

Appetite. Used for a psychlogical condition which gets manifested as a relish for something used usually in relation to food; controls amount of food intake in animals and is related with the energy requirements of body; depends upon colour, smell, taste, cooking medium, place of serving food etc. Controlled by staiety centres present in hypothalamus.

Appetitive Behaviour. The term used for describing the first phase of a series of actions undertaken by an animal after initial motivation towards a specific goal. For example, the appetitive behaviour of a hungry animal is that used in the search for food.

Applanate. Flattened horizontally.

Applied Ecology. The use of ecological concepts—*e.g.* succession, population growth and decay, and biogeochemical cycling—in issues directly relevant to human societies. Examples include studies of cereal ecosystems; forest ecology; pollution studies (*e.g.* of the reclamation of derelict land, of the disposal of toxic wastes, and of eutrophication and oligotrophication in rivers and lakes); advice on proposed changes in land use (*e.g.* the construction of tidal barrages); and biosphere conservation management.

Apple. The fleshy, pome, fruit of a tree of the genus Malus, or the tree which produces it.

Apposable. Capable of being arranged side by side or next to each other.

Apposition. The laying down of layers of cellulose on the inner surface of a plant cell wall to form the secondary cell wall. The process normally occurs once extension of the cell wall is completed and it serves to strengthen the overall cell structure. Compare intussusception.

Appositional Growth. Growth by deposition of material on previously formed layers, as in bone.

Apposition Image. When light is intense, the rays of light given out by some objects in front of ommatidia of compound eye (in insects etc.) penetrate through cornea. The oblique rays of eye will strike iris pigment and gets terminated there while straight rays of light reach upto focusing part of ommatidium and the image of only that part of the object gets formed from which the straight rays are passed thus separate images will be formed by each ommatidium. This is termed as apposition image and gets magnified and more clear (also known as mosaic vision).

Appressed. Lying close to or flat against.

Appressorium. A hyphal structure, formed by many parasitic fungi, that serves to effect penetration of the host epidermis. It is flattened and closely pressed to the outer surface of the epidermis. From the undersurface a narrow infection hypha or penetration tube is pushed through into the cell or cell spaces below. This then expands into a haustorium or develops into hyphae.

Approximate. Close together but not united.

Apricot. A tree producting a golden fruit with a stone, native to China. Apricots are grown in Europe and North-America. They belong to the plum family.

Apterium pl. Apteria. A bare or downy area between feather tracts of a bird.

Apteronotidae (class Osteichthyes, subclass Act-imopterygii, order Cypriniformes). Family of freshwater fish occurring in S. America. Members of the family possess a body tapering down to a narrow tail with a tiny tail fin. Related to the electric cells, the apteronotids can also produce weak electric charges.

Apterous. Lacking wings or winglike processes.

Apterygidae (kiwis; order Apterygiformes). Family of fowl-sized, brown-grey ratites. They have short, muscular legs, and a long flexible bill with nostrils at the tip. Their feathers are loose and hair-like. They have vestigial wings and no tail. They are small-eyed and mainly nocturnal, inhabiting thick forests and nesting in burrows. They feed on grubs and worms. The family comprises one genus (Apteryx) containing three species, all of them endemic to New Zealand.

Apterygota (class Insecta). The smaller subclass of the Insecta, which is now recognized as containing two orders :

Archaeognatha and Thysanura. Members of these orders are primitively wingless (in contrast to secondarily wingless insects, *e.g.* fleas, lice, etc.), terrestrial insects with ectognathous, biting mouthparts. The subclass formerly included the Protura, Collembola, and Diplura, which are now regarded as distinct classes of hexapod. Most Apterygota are free-living; elongate or oval; 3-15 mm long; and with direct larval development, moulting occurring throughout the life of the animal.

Apterygote. Any insect of the subclass Apterygota comprising primitive forms which undergo little or no metamorphosis, as bristletails and springtails. Sometimes proturans are included.

Apterylae. Area of bird skin in which feathers do not grow (Compare with pterylae).

Apus. A small freshwater crustacean which resembles the long-extinct trilobites in appearance and habits.

Aquarium

1. An artificial pond, pool, vessel, or container in which aquatic organisms are kept.
2. A place for the exhibition and care of aquatic organisms, especially fishes.

Aquatic. Pertaining to water, living in water *e.g.* fishes.

Aqueduct. A canal/channel which is used for conduction of fluid; very narrow mesocoel (cavity of mid brain) which connects diocoel (cavity of diencephalon) with metacoel (cavity of cerebellum), also known as iter/aqueduct of Sylvius.

Aqueduct of Sylvius. The cerebral aqueduct, a slender canal in the midbrain connecting the third and fourth ventricles.

Aqueous. Consisting principally of or of the nature of water.

Aqueous Humour. The fluid that fills the space between the cornea and the lens of the vertebrate eye. In addition to supplying the cornea and lens with nutrients, the aqueous humour helps to maintain the shape of the eye. It is produced and renewed every four hours by the ciliary body.

Aquifer. Body of rock, unconsolidated gravel, or sand stratum that is capable of storing significant quantities of water, that is underlain by impermeable material, and through which ground water moves. An unconfined aquifer is one in which the water table defines the upper water limit. A confined aquifer is sealed above and below by impermeable material. A perched aquifer is an unconfined groundwater body supported by a small impermeable or slowly permeable unit.

Aquifoliaceae. Family of trees or shrubs with leathery, alternate, evergreen, sometimes spiny, simple exstipulate leaves. They are usually dioecious. The flowers, borne in small cymes, are regular, tetra- or pentamerous. The petals and stamens are usually free. The fruit is a small drupe with three or more stones. Ilex aquifolium (holly), native in western and central Europe, is much grown for its foliage and for the scarlet fruits of the female trees. There are five genera, with more than 300 species, found throughout the world, except in arctic regions, but centred on the tropics.

Aquilegia (columbines; family Ranunculaceae). Genus of poisonous perennial herbs, which are usually erect, with alternate, two or three times ternate, compound leaves. The flowers are showy, regular, pentamerous, with petaloid sepals and long-spurred petals, and many free stamens and carpels which produce several-seeded follicles in fruit. Several species and hybrids are cultivated for their flowers. There are about 55 species, found mostly in the northern temperate zone.

Ara (macaws; family Psittacidae). Genus of medium to large parrots with long, graduated tails and huge bills. They are brilliantly coloured in red, blue, yellow, or green. They are frugivorous and found in rain forests, where they nest in holes in trees. There are 12 species, found in C. and S. America, and extinct species occurred in the W. Indies. A. macao (scarlet macaw) is widely kept in captivity.

Araban. (arabinan) A polysaccharide in which the major monosaccharide subunit is the pentose sugar arabinose. Arabans are found with pectic substances in mature primary cell walls and in hemicelluloses and gums.

Arabinose. A pentose sugar of the aldose group commonly found in plants. It often occurs in the polymerized form araban.

Araceae. Major family of monocotyledons, comprising tiny to giant herbs, and many bole climbers, which produce sap with irritant crystals. The leaves commonly have net-veins. Flowers are borne in a spike (spadix) subtended by an often colourful spathe; and are small, usually unisexual, the male apical, a perianth usually lacking. The fruit is a berry. There are about 110 genera, and about 2000 species, with pantropical distribution (especially America), with temperate outliers.

Arachis (family Leguminosae, subfamily Papilionatae). Genus of perennial herbs which are geocarpic. After flowering the inflorescence axis elongates so that the pods open below ground. A. hypogaea (peanut, groundnut, monkey nut) is not known in the wild, but is widely cultivated for the edible seed and oil; it is also used for soap and lubrication; the residues make cattle-feed, and the green haulms make excellent fodder. There are 15 species, occurring in America from the Amazon River south to Argentina.

Arachnid Member of the group of arthropods containing the spiders, harvestmen, scorpions and mites. Arachnids are most readily distinguished from insects by their four pairs of walking legs (an insect has three pairs). They have two pairs of appendages on the head: the first set are tiny pincers used for grasping and crushing prey, the second are palps used for grasping or sensing.

Arachnida. A class of terrestrial arthropods comprising about 65 000 species and including spiders, scorpions, harvest-men, ticks, and mites. An arachnid's body is divided into an anterior cephalothorax (prosoma) and a posterior abdomen (opisthosoma). The prosoma bears a pair of grasping or piercing appendages (the chelicerae), a pair of pedipalps used for manipulation or as sensory structures, and four pairs of walking legs. The opistho-soma may bear various sensory or silk-spinning append-ages (spinneret). Arachinds are generally carnivorous, feeding on the body fluids of their prey or secreting enzymes to digest prey externa ly. Spiders immobilize their prey with poison injected by the fanglike chelicerae, while scorpions grasp their prey in large clawed pedipalps and may poison it using the posteiror stinging organ. Uicks and some mites are parasitic but most arcahnids are free-living. Thep breathe either via tracheae (like insects) or by means of thin highly folded regions of the body wall called lung-books.

Arachnoid

1. Like a cobweb.
2. *annat.* Tne arachnoid membrane, the middle of the three meninges which enclose the brain and spinal cord.
3. *bot.* Covered with thin, loose hairs or fibers.

Arachnoid Membrane. One of the three membranes (meninges) that surround the brain and spinal cord of vertebrates. It lies between the pia mater and the dura mater. The arrachnoid membrane is very delicate and carries cerebrospinal fluid, which sustains and cushions the nervous tissue.

Araliaceae. Family of herbs, shrubs, climbers, and soft-stemmed pachycaul trees, which is the tropical counterpart to the Umbelliferae. The leaves are often large, and usually compound. The flowers are usually borne in compound umbels, and are small and regular, with four or five reduced sepals; five or three free petals and stamens; a disc; and the ovary inferior with five loculi and one, pendulous ovule. The fruit is a five-seeded drupe. There are about 56 genera, with about 700 species, most of them tropical.

Aramidae (limpkin; order Gruiformes) Monotypic genus (Aramus guarauna), a brown-olive bird, streaked with white, that is structurally crane-like and rail-like in appearance. The bill is long, laterally compressed, and decurved. The bird has a long neck, broad wings, a short tail, and long legs with long, unwebbed toes. It feeds on snails, nests on the ground or in bushes, and is found in wooded swamps in the southern USA, and in C. and S. America.

Araneae (spiders; class Arachnida). Very large and diverse order of predatory arachnids most of which feed on arthropods, although the large mygalomorphs will catch crestling birds and other vertebrates, some will catch tadpoles and small fish. The prosoma has a uniform dorsal carapace bearing 2—8 eyes. The chelicerae are large but have a fang (rather than a pincer) which bears the opening of a poison gland lying entirely within the chelicera in primitive species, or extending into the prosoma. The pedipalpi are leg-like and the complex structures of the male pedipalp are crucial in sperm transfer, the pedipalp introducing the sperm into the female epigyne (a sexual opening on the ventral surface of the abdomen) with a pumping action. The abdomen (opisthosonia) is usually unsegmented and attached to the prosoma by a narrow, stalk-like pedicel. The abdomen is extremely manoeuvrable, bears the silk spinning organs (spinnerets), and contains two pairs of respiratory organs (book lungs and posterior tracheae), the reproductive organs, and internal glands. Almost all species use their eight walking legs for running rapidly, and some jump by means of rapid rises in internal pressure. The legs have seven segments, the distal segment (pretarsus) having two comb-like claws and often a hook-like median claw for silk manipulation. Spider sensory apparatus varies from sensory setae, trichobothria, and slit sense organs to chemical and tactile receptors on the tarsi and pedipalpi. Vision is important and acute in the Salticidae (jumping spiders).

Aranediae (orb-web spiders; order Araneae, suborder Arancomorphae). Family of spiders which spin typical orb webs and whose legs are armed with many long spines. The chelicerae possess manv teeth, and many species of orb spinners have dwarf males. The orb web may be of complex construction, with viscid spirals and many radii. Prey caught on it is bitten, cut out of the web, and wrapped up. In many species, the spider may lie in wait at the hub or centre of the web; in others it may lie concealed close by, at the end of a taut signal thread. There are more than 2500 species, of which Araneus is the largest genus.

Araneus Diadematus (garden-cross spider, garden spider; suborder Araneomorphae, family Araneidae). Species of orb-spinning spider whose common name refers to its frequent occurrence in gardens and the patterns on its abdomen. The wells can be up to 40 cm in diameter, and can be found in woods and on heaths as well as in cultivated gardens, from late summer to autumn. The spiders have generally poor eyesight and may take prey items with strong warning coloration. Females are 12 mm long, and males 8 mm. This spider is widespread and common throughout Europe, and has been introduced into N. America.

Arapaima. A pike-like freshwater fish of South America. It is one of the largest freshwater fishes in the world, reaching a length of 2 metres or more and weighing over 100 kg.

Araucaria (family Araucariaceae). Genus of tall, columnar, coniferous trees, in which the crown is monopodial with radial limbs. The leaves are sharply pointed, crowded, and tough. The timber is valuable. A. bidwillii (bunya bunya) of Queensland produces edible seeds. A. angustifolia (parana pine) is a major timber tree in Brazil. Norfolk Island pin is A. heterophylla. Monkey puzzle (A. araucana), native to the S. Andes in Chile and Argentina, produces edible seeds. There are 18 species, of tropical to temperate forests in New Guinea, Queensland, Melanesia; two of the species are found in S. America, and 13 of them in New Caledonia.

Araucariaceae. Family of conifers comprising large trees which produce valuable timber. Several species are cultivated. The leaves are broad to lanceolate. The male cone is a catkin; the female is globose, and often massive, disintegrating at maturity. There are two genera (Agathis and Araucaria), native to the southern hemisphere (but with fossils in the northern hemisphere), and found mainly in Malesia and Melanesia, but also in S. America, and extending north of the equator in Malaysia.

Arboreal

1. Pertalning to or of the nature of trees.
2. Attached to or living in or among trees.

Arboreous. Heavily wooded : Arboreal, Arborescent.

Arborescent. Treelike in structure, appearance, or manner of growth.

Arboretum. An area devoted to the cultivation of a wide selection of all the woofy plants (trees, shrubs vines, etc.) that may be grown in a particular climatic region. Arboreta are maintained both as centres of research and as educational and recreational ares as.

Arborization. A treelike branching.

Arborviate. Refers to internal white matter present within cerebellum of brain; resembles a multibranched tree.

Arbovirus. One of a large heterogeneous group of RNA-containing viruses that are transmitted from animals to man through the bite of mosquitoes and ticks (*i.e.* arthropods, hence arthropod-borne viruses). Arboviruses cause various forms of encephalitis (inflammation of the brain) and serious fevers, such as dengue and yellow fever.

Arbor Vitae

1. *bot.* An evergreen tree of the genus Thuja, a common ornamental shrub.
2. *anat.* A treelike structure composed of white matter seen in a median section of the cerebellum.

Arbuscula. A small shrub having the form of a tree.

Archaeoceti (ancient whales; cohort Mutica, order Cetacea). Extinct suborder comprising the oldest and most primitive cetaceans which flourished in the Eocene and which may have originated in Africa. They were comparable in size with modern porpoises, had an elongated snout, and nostrils on top of the skull. The braincase was long and low. The front teeth were peg-like, the cheek teeth heterodont and characteristic of primitive carnivores. There were 44 teeth in all. The hind legs were reduced to vestiges which did not protrude from the body. They were fish-eating carnivores that had adopted an aquatic life to which they were more highly adapted than, for example, modern seals.

Archaeocyathida (class Anthocyanthea). Order of Lower and Middle Cambrian Archaeocyatha. Some were solitary, others colonial. They have a cylindrical, conical, or discoid cup, often of irregular outline or with outgrowths. The outer wall has simple pores or is partially displaced by dissepiments. The inner wall usually has one longitudinal row of pores between intercepts. The intervallum contains dissepiments, rods, and bars or septa. The order contains two suborders and 19 families.

Archaeogastropoda. Group comprising what may be the most primitive gastropods. Members of the class Gastropoda may be subdivided according to their respiratory apparatus. In the Archaeogastropoda this seems to be in the most primitive condition and consists of just two gills. Patella vulgata (the limpet) belongs to this group.

Archaeocyathea. Extinct group of reef-forming organisms known only from the Cambrian. In some respects they were similar to both spongers and corals, and may represent a truc advance in evolutionary grade over the former. It is possible that they lived in a symbiotic relatonship with some trilobites. The cause of their extinction is not known.

Archaeognatha (class Insecta, subclass Apterygota. One of the two orders of the Apterygota, comprising fusiform, subcylindrical insects which can run fast and jump by flexion of the abdomen. The body is covered with pigmented scales, and the thorax is strongly arched. Ocelli are present, and the compound eyey are large and contiguous. The mandibles are monocondylar (with a single articulation), and the maxillary palps are long and seven-segmented. The abdomen has ventral styles on segments 2—9, used for running on sloping ground, and the posterior portion of the abdomen bears a pair of cerci and a single, much longer, appendix dorsalis. These insects are free-living and nocturnal, feeding on lichen, algae, and vegetable debris. Post-embryonic development is slow (six months to two years), and moulting occurs throughout the entire life-span. The 250 or so species are of little economic importance.

Archaeopteryx. A fossil brid of the genus Archaeopteryx with teeth embedded in sockets, separate metacarpal bones, and a lizardlike tail with feathers along the sides; of the late Jurassic.

Archaeornis. A fossil bird of the genus Archaeornis similar to Archaeopteryx.

Archaezoic. A geological era that started after formation of earth crest about 3700 million years ago and had duration/ age of 2100 million years. Protista (prokaryotic animals) were evolved during this period.

Archegonial. Of, pertaining to, or bearing archegonia.

Archegoniophore. An upright structure consisting of a stalk and cap that bears the archegonia in certain liverworts. In Marchantia sterile raylike structures radiate form the cap giving it the appearance of a chimney sweep's brush. In Reboulia the archegoniophore is umbrella shaped. The archegonia are borne in groups on the lower surface of the cap, the groups being separated by involucral scales termed perichaetia. The stalk of the archegoniophore does not elongate until fertilization has occurred. Thus the antherozoids can swim to the archegonia before the water film between the thallus and archegoniophore is disrupted. Compare antheridiophore.

Archegonium. The multicellular flaskshaped female sex organ of bryophytes, pteridophytes, and many gymnosperms. Such plants are described as archegoniate to distinguish them from algae, which do not possess archegonia. The dilated base, the venter, contains the oosphere (female gamete). The cells of the narrow neck liquefy to allow the male gametes to swim towards the oosphere. The archegonium is thus an adaptation to the terrestrial environment as it provides a means for the male gametes to reach the female gamete. Compare antheridum.

Archenephros. Used for describing a type of kidney which is present in most primitive vertebrates; having one nephric tubule of each side per body segment and all the tubules joining into a longitudinal duct which is opening into cloaca. Also known as holonephros.

Archenteron (gastrocoel). A cavity within an animal embryo at the gastrula stage of development. All or part of the archenteron eventually forms the cavity of the gut. It is connected to the outside by an (opening the blastopore), which becomes either the mouth, the mouth and anus, or the anal opening of the animal.

Archoecyte. A large amebocyte in the mesenchyme of sponges.

Archeology. In human evolution, the study of objects of cultural significance, as tools, weapons, ornaments, and utensils.

Archeozoic Era. The earlist period of geological time, with an estimated duration of two billon years. Fossil remains are lacking, but there is evidence that primitive life forms existed.

Archeptrygium. A type of fish fin having a single segmented axial rod from sides of which numerous secondary segmented radials arise; present in Neoceratodus (a lung-fish).

Archerfish. A small fish of muddy coastal waters and swamps in India and South-East Asia. Archerfishes have a remarkable way of obtaining insect food : they squirt a stream of water droplets at insects resting on overhanging plants and knock them into the water. Archerfishes have been known to hit insects from a distance of 2 metres.

Archesporium. Refers to the single cell or group of plant cells in the sporophyte from which spores may eventually develop in a sporangium.

Archetype. A basic model, plan, or prototype; a generalized type of structure assumed to have been possessed by ancestors of a group.

Archilochus (hummingbirds; family Trochilidae). Genus of small hummingbirds with green backs and white under-parts. The males have forked tails, the females have blunt tails with white spots. A. colubris (ruby-throated hummingbird) is a long-distance migrant from Canada to Central America. There are two species, both of which breed in N. America and winter in Central America.

Archinephros. A holonephros, a primitive type of kidney present in the larva of certain hagfishes and amphibians.

Archipallium. The older, primitive portion of the cerebral cortex which, in mammals, comprises the hippocampal formation, a part of the rhinencephalon. Compare neopallium.

Arcoida (phylum Mollusca, class Pelecypoda). Order of sedentary, epifaunal pelecypods in which the adults are byssate or free-living. Typically they are isomyarian, taxodont, and equivalved. The shells are normally trapezoid, and composed of crossed-lamellar calcium carbonate. They have filibranchiate gills, and flat, cardinal areas on the dorsal margin. They first appeared in the Lower Ordovician.

Arctic. Pertaining to, or of the region of the North Pole; polar; frigid.

Archontophoenix (family Palmae). Genus of palms, in which the trunk is solitary, and hoop-marked; and the leaves pinnate, with a conspicuous crown-shaft, below which the much-branched pendulous inflorescences appear. There are two species, native to subtropical E. Australia, and widely planted as stately ornamentals. A alexandrae (the king palm) is of very similar general appearance to Dictyosperma album.

Arbacioida (superorder Echinacea). Order of regular echinoids in which the imperforate, non-crenulate primary tubercles are poorly developed. Adventitious skeletal material on the test surface, simulating tubercles, is commonly developed, but does not bear spines. They first appeared in the Middle Jurassic.

Archiannelida. A class of the Annelida having small marine worms most of which are scavengers with a protrusible tongue for transferring food to the mouth.

Arctic Fox. A small fox with short ears and muzzle, giving it a somewhat catlike appearance. It lives in the far north and its greyish yellow summer coat turns white in winter. Like the polar bear, the arctic box has hairy soles on its feet. This helps it to get a grip on the snow and ice.

Arctic Hare. A large hare found on the treeless tundras of North America and in Greenland. In the northernmost part of its range, the arctic hare is white all year round; further south it has a grey back in summer; only the tips of the ears are always black.

Arctic Scrub. Scrub composed of plants that average about 60 cm high, and including dwarf willows, birches, or both. It occurs in damp hollows and along the edge of water in the low-arctic belt of the tundra. Northwards it becomes more stunted and limited in area, and more or less disappears beyond the centre of the middle-arctic belt.

Arctidae (tiger moths, ermines, footmen, tussocks, woolly bears; order Lepidoptera). Family of small to medium sized moths, many of which are brightly coloured; and which are often distasteful to predators. The larvae are often hairy (and so called 'woolly bears'), and hairs are used to construct the cocoon. Arctia caja (garden tiger moth) is widespread in Europe as far north as Lapland. There are about 10,000 species. The family has a world-wide distribution.

Arctocyonidae (superorder Protoungulata, order Condylarthra). Extinct family of unspecialized primitive placental mammals with more than 20 genera, which lived from the Cretaceous to the Eocene but which were most abundant in the Palaeocene. Some were the size of modern bears, and were probably omnivorous; others were smaller and herbivorous. The body was slim, the limbs slender, the feet clawed. The braincase was small, and the teeth primitive, with moderately sharp cusps. Protoungulatum of the N. American Upper Cretaceous is the oldest known ungulate.

Arcualia. Two pairs of cartilaginous structures (basidorsals and basiventrals) which develop in the mesodermal sheath surrounding the notochord in vertebrates. They give rise to the neural and hemal arches.

Arcuate. Curved like a bow.

Arecidae. A subclass of the monocotyledons containing both herbaceous and arborescent (the palms) plants. Members of the Arecidae differ from other monocotyledons in having broad net-veined petiolate leaves. Their numerous small flowers are usually unisexual and grouped into an inflorescence subtended by a spathe. Four orders are commonly recognized : the Arecales, which contains one family, the Arecaceae or Palmae (palms); the Cyclanthales, which also contains one family the Cyclanthaceae; the Arales, which contains the two families Lemnaceae (which includes the duckweeds), and Araceae (which includes the arums or aroids); and the Pandanales, which comprises the one family Pandanaceae (which includes the screw pines). In some classifications the order Typhales is placed in the Arecidae, while in orders it is allocated to the Commelinidae. The Typhales includes two families, the monotypic Typhaceae, containing the cattails and reedmace bulrush (Typhus), and the monotypic Sparganiaceae, containing the bur-reeds (Sparganum).

Areca (family Palmae). Genus of palms, which grow solitarily, or in small clumps. The leaves are pinnate, with a conspicuous crown-shaft. Inflorescences occur distally with only male flowers, and near the base triads a big female is flanked by two males. The fruits have a fleshy mesocarp. The betel-nut comes from A. catechu. There are about 50 species, native to tropical Asia.

Area Centralis. Refers to a small depression present towards the bosterior side of retina (where there are more cones than rods) and its centre is exclusive region of cones termed as fovea centralis and in primates area centralis possesses yellow-pigment thus termed as macula lutea.

Area Pellucida. Refers to the central area/circle in the blastodise of reptiles and birds over blastocoel which is less opaque and also called area opeca. Constituent of opiblast and the region where entire chick gets developed.

Arenaceous. Sandy; consisting of, or of the nature of sand; growing in sand.

Arenaria

1. (family Caryophyllaceae). Genus of low herbs or dwarf shrubs which have slender shoots, tiny, opposite leaves, and small, usually white flowers with five undivided petals, ten stamens, and three styles (rarely more). The fruit is a tiny capsule, usually six-toothed on opening. Several mat-forming species are cultivated as rock-garden plants (sandworts). There are about 160 species, cosmopolitan in distribution, but most common in the northern temperate zone to the arctic.
2. (turnstones; family Charadriidae). Genus of shore birds including A. interpres (turnstone), which breeds in the artic and migrates to the S. hemisphere in winter. It uses its short black beak to turn over pebbles and molluscs in the search for food.

Arenga (family Palmae). Genus of palms found in their native habitat as treelets to stout trees. The leaves are pinnate, with pinnate nervation, the sheath coarsely fibrous. Inflorescences often develop basipetally, paired male flowers falling before the solitary female develop. The fruits have a soft irritant mesocarp and 1-3 seeds. The sugar palm is A. pinnata. There are about 20 species, native from the Himalayas, and found from Hainan to New Guinea.

Arenicolous. Growing in sand or sandy areas; living in or on sand.

Arenose. Sandy; gritty.

Areola

1. The colored ring surrounding a nipple.
2. The part of the iris which encircles the pupil.
3. A minute space in a tissue.
4. A bare space between scales on the feet of birds.
5. A median, longitudinal strip on the dorsum of a crayfish.

Areolar Tissue. Refers to a type of loose connective tissue which is present around visceral organs, among muscles, around blood vessels, around nerve fibres and structural element of atorma of various organs. It is also forming basement membrane of epithelia. It has jelly-like matrix with elastic, collagen and reticular fibrils. There are fibroblasts (help in healing), histeocytes (phagocytic), plasma cells (antibody forming), mast cells (heparin producing) and lymphocytes (antibody forming) in this tissue.

Areole

1. A sunken cushion representing a condensed lateral shoot from which spines, branches, and flowers arise in cacti. Areoles may occur either singly on tubercles (*e.g.* in Mammillaria zeilmanniana) or in rows along raised ridges.

2. Any area outlined on a surface, *e.g.* a segment of leaf lamina surrounded by veins. A surface divided into areoles, *e.g.* lichen thalli split by cracks into roughly hexagonal areas, is described as areolate.

Argali. The largest of the wild sheep, the argali stands well over one metre at the shoulder and has spirally twisted horns up to 2 metres long. It is found in central Asia.

Argasid. Any acarid of the family Argasidae which includes the soft ticks, as Ornithodoros, transmitter of the infective agents of tickbite fever, Q fever, tularemia, and relapsing fever.

Argentaffin. Having an affinity for silver or chromium salts, as argentaffin cells of the alimentary canal.

Argentinidae (herring smelts, argentines; class Osteichthyes, subclass Actinopterygii, order Salmoniformes). Family of small, marine, large-eyed, silvery fish that are found in coastal areas of the oceans. Preferring deeper water, schools of the Atlantic argentine, Argentina silus, can be caught in large numbers and then used for fishmeal. There are about 12 species.

Arginine. A basic amino acid with the formula $H_2NC(:NH)NH(CH_2)_3CH(NH_2)COOH$ (illustration at amino acid). It is important in histone proteins, which are rich in this amino acid.

Argonaut. A marine mollusc related to the cuttlefish and living in warm waters. The female, up to 30 cm long, lays her eggs in a water-thin shell. The male, which rarely reaches 1 cm in length, was identified by zoologists little more than 100 years ago.

Argyroneta Aquatica (water spider; suborder Araneomorphae, family Agelenidae). The only species of spider to live permanently below water, where it constructs an inverted retreat filled with air carried from the surface by means of its rear legs. The carapace is pale to dark yellowish-brown, with darker striations, and the abdomen, which has a grey pubescence, is covered with an air film, giving a silvery appearance. Females are 8-15 mm long, and males 9-12 mm. The spiders feed mostly an isopods, which are consumed within the air bell. Special bells are made for the sperm web and eggs, and the adults over-winter in the bell, which is specially strengthened. The hatchlings remain within the retreat until after the fourth moult, when they are well covered with pubescence.

In humans, the bite of Argyroneta can cause itching, shivering, vomiting, and a slight fever lasting 2-3 days. The species is widespread in the Palaearctic region.

Argyrophilic. Readily impregnated or stained with silver.

Arid. Dry; barren; lacking in life.

Ariidae (Tachysuridae; sea catfish; class Osteichthyes, subclass Actinopterygii, order Siluriformes). Family of mainly marine catfish with a forked tail-fin, adipose dorsal fin, naked skin, and several pairs of barbels around the mouth. They are found in tropical and subtropical seas around the world. Many species, including the gaff-topsail catfish, Bagre marinus, carry out oral incubation, with the males incubating fertilized eggs in the mouth for up to six weeks. There are many species.

Aril. An outgrowth that grows around and may completely enclose the testa (seed coat) of a seed. It develops from the placenta, funicle, or micropyle of an ovule. The aril surrounding the nutmeg seed forms the spice mace.

Arillate. Having an aril.

Arillode. A false aril originating from the funicle or chalaza, as in the nutmeg.

Arista. A bristlelike appendage, as *a.* that on the antenna of certain dipterous insects; *b.* the awn of cereal plants.

Aristate. Describing a structure, such as a glume or lemma, bearing an awn. Compare bearded.

Aristogenesis. Theory, like those of entelechy, nomogenesis, and orthogenesis, that claims that evolution proceeds along a determined path. Today, however, most people accept that natural selection does not direct evolution towards any particular kind of organism or physiological attribute, and that there is no mysterious inner guiding force either.

Aristolochiaceae (birthworts). Family of climbers, and a few herbs. The flowers are usually zygomorphic, often foetid; have a calyx that is often enlarged and highly coloured; no corolla; six to many stamens; and the ovary inferior and 4-6 locular. The capsule contains many small seeds. The roots are often medicinal. Aristolochia is Dutchmen's pipe. There are seven genera, with about 625 species, of warm-temperate to tropical distribution.

Aristole's Lantern. A complicated, toothed structure consisting of several ossicles surrounding the mouth of a sea urchin.

Arixeniidae (class Insecta, order Dermaptera). Family of earwigs which are hairy, wingless, and robust, with short, rod-like forceps. Only two species are known, and they are associated with two species of bats found in Malaysia, Indonesia, and the Philippines.

Arm

1. In man, the portion of the upper extremity from shoulder to elbow.
2. In common usage, the portion from shoulder to wrist.
3. A structure of a corresponding nature, as *a.* the anterior limb of a vertebrate; *b.* the ray of a sea star; *c.* a tentacle of an octopus; *d.* an extension of the manubrium of a medusa.

Armadillidae (pill bugs; subclass Malacostraca, order Isopoda). Family of terrestrial pill bugs which have five pairs of tracheal gills and lack a water-transport system. All members of the family can roll up completely in response to disturbance or in the presence of predators.

Armadillidium Vulgare (pill bug; subclass Malacostraca, order Isopoda). Widely distributed and abundant species of isopod which has a rectangular to ovoid shape. Individuals may be coloured grey, black, or pale yellow, and can roll up completely when disturbed, assuming a spherical appearance. They may be up to 6 mm in diameter when rolled up, and are found on the upper shore, beneath stones. Parthenogenetic females may occur in bisexual races.

Armadillo. A group of armoured omnivorous animals ranging from Patogonia to the south-western USA. Largest is the giand armadillo of South America which reaches one one metre in length. The three-banded armadillo is the only one able to roll up into a protective ball. Other

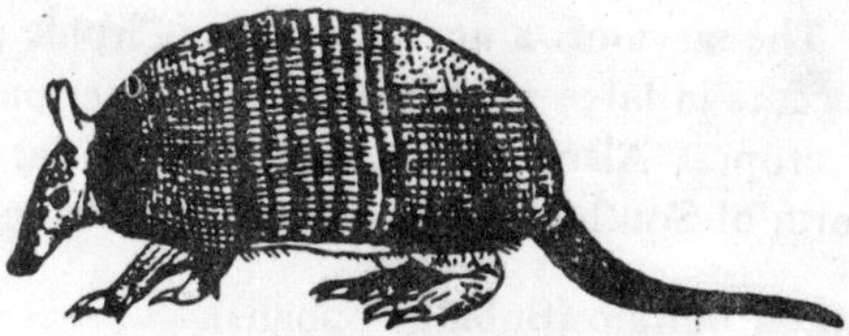

Armadillo

species include the ninebanded armadillo of Mexico and the fairy armadillo of the Argentinian plains.

Armature. Collectively, defensive and protective structures, as teeth, horns, spines, or thorns.

Armed. Bearing hooks, with reference to the scolex of tape-worms.

Armored. Possessing a protective covering of heavy scales or bony plates.

Armored Scales. Scale insects of the family Diaspididae whose bodies are covered with a hard, waxy scale, as the San Jose scale.

Armoured Catfish. Freshwater South American fishes with whisker-like barbels and bony plates protecting part or all of the body. Armoured catifishes can live out of water for a considerable time. They can even travel overland from one river to another, dragging themselves along with their strong pectoral fins.

Army Ant. Also called driver ants or legionary ants, these insects are confined to the warmer parts of the world and have no permanent nests. The colonies move about in long columns, with large-jawed workers on the outside to defend them. Any animal which is unable to move out of the way of the column is eaten, no matter what its size. Males have wings and look very different from the other members of the colony. Their light brown bodies may exceed 2.5 cm in length and they are known as sausage flies.

Armyworm. The larva of a noctuid moth (Cirphis unipuncta) which occurs in large numbers and is a serious pest to cereal crops. Also Laphygma frugiperda, the fall armyworm of Southern states, a pest to corn and cotton.

Arna. The water buffalo (Bubalus bubalis).

Arolium. In insects, a padlike structure at the tip of the tarsus.

Arousal. A level of physiological and behavioural responsiveness in an animal, which tends to vary between sleep and full alertness. It is controlled by a particular part of the brain (the reticular activating system) and can be detected by changes in brain electrical activity, heart rate and muscle tone, responsiveness to new stimuli, and general activity.

Arrector. An erector muscle, especially the arrector pili, consisting of smooth muscle fibers attached to the base of a hair follicle.

Arrector Pilli. A small bundle of unstriped muscle fibres which are attached to hair whose contraction makes hair almost vertically erect to the surface of skin.

Arrnenotoky. Female parthenogenesis in which only males are produced.

Arripidae (Australian salmon; class Osteichthyes, subclass Actinopterygii, order Perciformes). Small family confined to the temperate waters of Australia and New

Zealand, comprising two species of fish which are often abundant along the coastal areas, and of considerable economic importance, The Australian salmon, Arripis trutta (not related to the true salmon), is a large carnivorous fish, often exceeding 5 kg in weight. It is highly regarded as a sporting fish, but commercial catches usually go to canneries.

Arrowhead. An aquatic plant growing in quiet muddy waters. It has white flowers, arrow-shaped leaves and edible tubers.

Arrowroot. A tropical West Indian herb. Starch from its tuberous rhizomes makes an easily-digested food.

Arrow Poison Frog. One of the most powerful poisons in nature is contained in the skin of these South and Central American frogs. It has long been used by Indians for making poison-tipped arrows. Most species of arrow poison frog have bright colours which act as warning signals to would-be predators.

Arrow Worm. A group of long, transparent, worm-like animals that are abundant in marine plankton. They are most common in warm seas but a few species are found in the seas around NW Europe. Their bodies are often arrow-shaped and there are fins on the sides and tail.

Artamidae (wood-swallows; order Passeriformes). Family of smallish, brown or grey birds with black and white markings and long, painted wings. They are the only passerines with powder-down feathers. They are highly social; and insectivorous, mainly aerial feeders. They inhabit clearings and open country near water, nesting on ledges or in trees. There is one genus (Artamus) containing 10 species, found in India, S. China, SE Asia, Australasia, and some S. Pacific islands.

Artemisia (family Compositae). Genus of perennial (or, rarely, annual) herbs or low shrubs, often aromatic, with leaves that are alternate, and much divided pinnately into narrow segments. The flower heads are tiny, usually numerous, often woolly, and gathered into racemes or panicles. The receptacle is flat, and naked; all the florets are tubular, surrounded by overlapping, scarious-edged bracts. There is no pappus. The genus includes mugworts, and wormwoods. Wormwood (A. absinthum) is used for flavouring beers and other drinks. There are about 200 species, occurring mostly in dry grasslands of the northern temperate semi-arid, and arctic zones.

Arteriole. A small muscular blood vessel that receives blood from the arteries and carries it to the capillaries.

Arterio-venous Anastomosis. A small muscular blood vessel that is carrying blood directly from the arterioles to the venules and bypasses the capillary network. It is stimulated by sympathetic nerves.

Artery. A blood vessel that carries blood away from the heart towards the other body tissues. Most arteries carry oxygenated blood (the pulmonary artery is an exception). The large arreries branch to form smaller ones, which in turn branch into arterioles. All arteries have muscular walls, whose contraction aids in pumping blood around the body. The accumulation of fatty deposits in the walls of the arteries leads to atherosclerosis, which limits and may eventually block the flow of blood. Compare vein.

Arthrobranch. A gill which arises from the membrane joining an appendage to the body, as in the crayfish.

Arthrodia. A joint which permits gliding or sliding movement as that between articular processes of the vertebrae.

Arthrodire. An extinct fish of the order Arthrodires with bony plates covering the anterior portion of the body.

Arthromere. A somite or metamere, a segment of a segmented animal.

Arthrophyte. A plant of the division Arthrophyta which includes the horsetails, characterized by a jointed stem; an articulate.

Arthropod. A member of the largest animal phylum, the Arthropoda. All arthropods have jointed limbs and are normally covered by a stout cuticle. The latter usually contains a horny protein called Chitin. Like the annelid worms, from which they are probably descended, the arthropods show definite segmentation of the body. They include Insects, Crustaceans, Arachnids, Bristletails, Centipedes and Millipedes.

Arthropoda. A phylum of invertebrate animals comprising over one million species—the largest in the animal kingdom. Arthropods inhabit marine, freshwater and terrestrial habitats worldwide. Characteristically, they possess an outer body layer—the cuticle—that functions as a rigid protective exoskeleton; growth is thus possible only by periodic moults (ecdysis). The arthropod body is composed of segments (metameric segmentation) usually forming distinct specialized body regions, *e.g.*, head, thorax, and abdomen. These segments may possess hardened jointed appendages, modified variously as mouthparts, limbs, wings, reproductive organs, or sense organs. The main body cavity, containing the internal organs, is a continuous blood-filled haemocoel, within which lies the heart. The origins and relationships of the various groups of arthropods remain uncertain. The major classes are : Arachnida (spiders, scorpions, mites, and ticks), Crustacea (shrimps, barnacles, crabs, etc.), Insecta (insects), Myriapoda (centipedes and millipedes), and the wormlike Onychophora (*e.g*, Peripatus).

Arthrospore. A spore produced by certain fungi by fragmentation of the hyphae, as in some actinomycetes.

Artichoke, Jerusalem. A tall coarse, sunflower-like plant up to 3.5 metres tall, native to North America. Its knobbly tubers are eaten.

Artichoke, Globe. A grey-green, thistle-like plant up to 90 cm tall. The fleshy scales of the flower-heads are eaten.

Article

1. *bot.* A segment of a fruit of pod, as in fegumes.
2. *zool.* A serrated process at the tip of the chelicera of a tick.

Articulamentum. The inner calcareous layer of the plate of a chiton.

Articular. A small bone of the lower jaw in bony fish (Osteichthyes), amphibians, and reptiles that makes a hinge joint with the quadrate bone of the upper jaw.

Articular Bone. A cartilage bone in lower vertebrates which forms the posterior portion of the lower jaw articulating with the quadrate.

Articular Cartilage. A thin layer of hyaline cartilage which covers the end of a bone in a diarthrodial joint.

Articulata

1. (Phylum Brachiopoda). Class of brachiopods, with calcareous valves which are impunctate, punctate, or pseudopunctate. The fibrous or prismatic secondary layer and non-fibrous primary layer are well differentiated. The periostracum is of protein. Hinge teeth and dental sockets are developed. Lophophore support is formed from modifications of socket ridges to give crura, loops, or spires. The alimentary canal ends blindly. They first appeared in the Lower Cambrian. There are six orders, with 37 superfamilies.

2. (sea-lilies; subphylum Crinozoa, class Crinoidea). Subclass of sea-lilies (crinoids) in which the basal plates are small or strongly reduced. Articulation between the radial and brachials, and in the majority of brachials, is muscular, with a well developed fulcral ridge. The arms are always uniserial. All post-Palaeozoic crinoids belong to the Articulata.

[1]Articulate

1. Jointed; formed of segments, united by joints.
2. Distinct, clear; formed of words or syllables, as articulate speech.

[2]Articulate

1. *zool.* A brachiopod with a hinged shell.
2. *bot.* An arthrophyte.

[3]Articulate. *anat.* To form a joint.

Articulation

1. *zool.* (a) A joint between two bones or between a bone and cartilage.

   (b) A point of union between any two parts or segments of an animal.

2. *bot.* (a) A point of union between two separable parts.

   (b) A node or a joint.

Articulator. A movable structure, as the lips, teeth, tongue, or soft palate, which functions in the production of distinct sounds in articulate speech.

Artifact

1. A structure seen in fixed preparations of cells and tissues which is not present in real life.
2. An object, as a tool or utensil, indicating human workmanship.

Artificial. In ecology, applied to a community, or habitat, or both that have been changed drastically from their original condition by human activity so that they are completely different from and unrelated to the community or habitat originally present at the site. Examples include the

Norfolk Broads (England), formed by the natural flooding of medieval peat cuttings; man-made reservoirs; and the reclamation of land from the sea, as in the polderlands of Holland.

Artificial Insemination. The deposition of semen, using a syringe, at the mouth of the uterus to make conception possible. It is used for selected breeding of domestic animals and in humans in some cases of importence and infertility. It is timed to coincide with ovulation in the female.

Artificial Propagation of Fish. Type of fish culture for stocking purposes. It involves 'stripping' eggs of ripe female fish followed by mixing with milt from mature males. The fertilized eggs are then either distributed at once or incubated and hatched in special receptacles before distribution to waters that have to be stocked. Increasing numbers of fish species are being propagated in this way.

Artificial Selection. Selection by humans of individual plants or animals from which to breed the next generation, because these possess the most marked development of the required attributes. Typically the process is repeated in successive generations until those attributes are fixed in the descendent offspring. Such artificial selection can result in dramatic changes, such as took place in the domestication of plants and animals from their wild forebears.

Artiodactyl

1. An even-toed mammal.
2. Any member of the order Artiodactyla, even-toed mammals, which include the pigs, peccaries, hippopotamuses, and all ruminants.

Artiodactyla. An order of hooved mammals comprising the even-toed ungulates, in which the third and fourth digits are equally developed and bear the weight of the body. The order includes cattle and other ruminants (Ruminantia), camels, hippopotamuses, and pigs. All except the latter are herbivorous, having an elongated gut and teeth with enamel ridges for grinding tough grasses. Compare Perissodactyla.

Artocarpus (family Moraceae). Genus of small to medium trees, yielding latex from the bark, and useful timber. The leaves have big stipules, commonly divided. Male flowers are borne in catkins, female flowers in heads. The fruit is multiple, formed on the fleshy receptacle by concrescent achenes surrounded by a fleshy perianth, often cauliflorous, and gigantic, in A. integrifolia (jak or jackfruit) attaining 1×0.6 m. The fruit of A. altilis (breadfruit) is starchy, and bread-like when roasted. There are 47 species, occurring in tropical SE Asia.

Arum. The name given to a family of mainly tropical plants in which the poker-like flower-spike or spadix is surrounded by a large sheath or spathe. Many give off unpleasant smells which attract pollinating flies. A number have sword-shaped leaves. The common arum or cuckoo-pint is a wild variety with poisonous red berries.

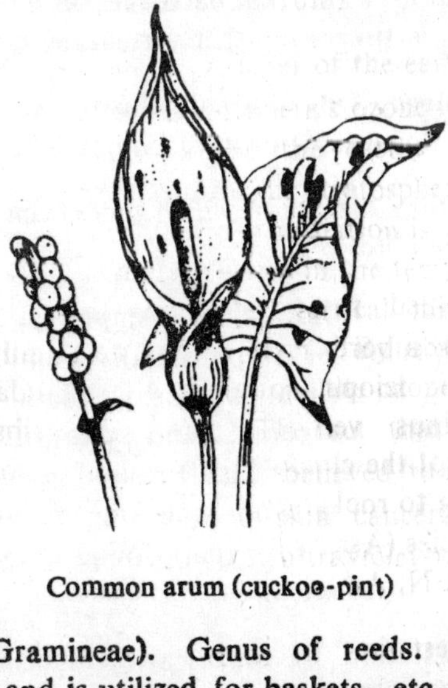

Common arum (cuckoo-pint)

Arundo (family Gramineae). Genus of reeds. A. donax may attain 7 m and is utilized for baskets, etc., and formerly for arrow shafts. It is good for making paper. Gardeners' garters is a variegated medium to large, elongate, bristly flies, which have a horny proboscis adapted for piercing. The wing vein R_1 is very long, and M_1 terminates some distance beyond the apex of the wing. The legs are powerful aad flexible, with large pulvilli, and a bristle-like empodium. A prominent tuft of hairs, or 'beard', and protruding eyes are also characteristic of the asilids. Virtually all adults are predators of other insects, using their long legs to grasp and carry off prey. The proboscis is sharp, points forward, and is used to pierce the prey and suck it dry. It can easily enter the hardest of cuticles, and in certain cases may inject venom. Asilid larvae are either predacious or scavengers in soil, leaf-litter, or other ground substrates. Larvae are amphineustic, and their mouthparts comprise a hook-shaped labrum, sharp mandibles, and broad mavillae with two-segmented palps. The Asilidae is the largest family within its suborder, with at least 4000 described species.

Arytenoid. Referring to a pair of laryngeal cartilages which are lying posterior to thyroid cartilages in mammals; anterior to cricoid cartilages in other vertebrates.

Arytenoid Cartilage. One of a pair of cartilages in the larynx to which the vocal folds or cords are attached.

Asbestosis. A lung disease in which there occurs the formation of asbestos fibres.

Ascaphidae (tailed frogs; class Amphibia, order Anura (Salientia), suborder Amphicoela). Family of amphibians that, like leiopelmatids (Leiopelmatidae), have nine amphicoelous vertebrae, and free ribs. A muscular extension of the cloaca ('tail') is used in copulation. Tadpoles cling to rocks with a sucking mouth. There is a single species (Ascaphus truei), found in mountain streams in western N. America.

Ascariasis Intestation by an ascarid, especially Ascaris, a parasitic nematode inhabiting the intestine.

Ascarid. Any nematode of the family Ascaridae to which Ascaris lumbricoides, a human intestinal parasite, belongs.

Aschelminthes. Animals of organ grade of organisation, triploblastic having pseudocoel and alimentary canal, *e.g.*, nematodes like Ascaris, Enterobius etc.

Asclepiadaceae. Family of erect or twining shrubs or perennial herbs, sometimes fleshy, and usually with white sap. Leaves are opposite, or whorled. Inflorescences are cymose. The flowers are regular, and pentamerous, the petals contorted, stamens often with two pollinia; the ovary semi-inferior. The fruit consists of paired follicles. The seeds are tiny, and plumed. There are about 130 genera, with 2000 species, mainly of the tropics and subtropics, especially of S. America.

Ascending

1. Rising, going upward.
2. *bot.* Rising obliquely upward from a prostrate position.

Ascending Series. A series of animals or plants arranged in order of complexity from the simplest to the most complex.

Aschelminth. Any member of the phylum Aschelminthes, pseudocoelomate invertebrates comprising the rotifers, gastrotrichs, kinorhynchs, priapulids, nematodes, and nematomorphs.

Ascidian. A sea squirt, a sessile tunicate of the class Ascidiacea.

Ascidium. A flask- or pitcher-shaped structure, as the leaf of the pitcher plant.

Ascocarp. The fruiting body of all ascomycete fungi except the Hemiascomycetes. It consists of an aggregation of hyphae surrounding the asci. The various types of ascocarp are used in dividing the Ascomycotina into classes.

Ascoceratida (class Cephalopoda, subclass Nautiloidea). Order of orthoconic and cyrtoconic cephalopods that have a thin juvenile portion and an inflated adult portion to the test. The siphuncle is near-central. Siphuncular and cameral deposits are not found. The order contains only 13 genera, and all are uncommon fossils. They range in age from Middle Ordovician to Upper Silurian.

Ascogenous. Producing or giving rise to asci.

Ascogenous Hypha. Any of a number of small multinucleate branches that develop from an ascogonium. Ascogenous hyphae give rise to asci by crozier formation.

Ascogonium. The female gametangium (ascogonium) of Ascomycetes.

Ascomycete. Any fungus of the class Ascomycetes which includes the sac fungi, characterized by the production of ascospores in a distinctive saclike structure, the ascus. Common representatives are the yeasts, blue and green molds, powdery mildews, ergot, cup fungi, sponge mushrooms, and lichens.

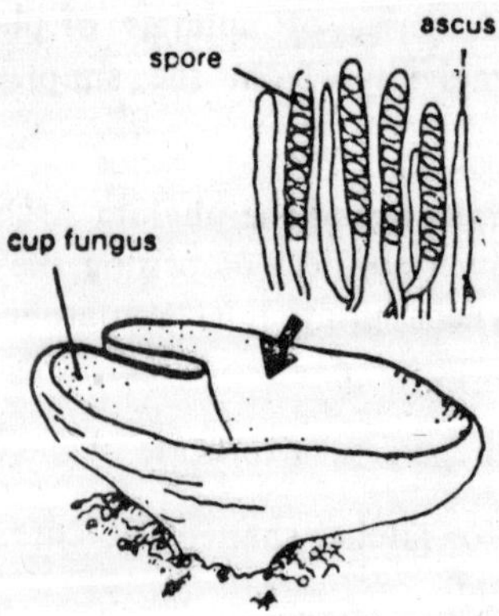

Ascomycete

Ascomycotina (sac fungi). A subdivision of the Eumycota, most of whose members have a mycelial septate thallus though some (the yeasts) are unicellular. It contains about 16500 species in almost 2000 genera. Sexual reproduction is by ascospores, with eight characteristically developing within an ascus. It is divided into the classes. Hemiascomycetes; Plectomycetes Pyrenomycetes; Discomycetes; Laboulbeniomycetes, which are ectoparasites of arthropods; and Loculoascomycetes. The sac fungi are often placed in the class Ascomycetes, and divided into the two subclasses Hemiascomycetidae (or Protoascomycetidae) and Euascomycetidae. About half the ascomycetes have developed a special mode of nutrition in which algal cells are incorporated in the thallus.

Ascon. A type of sponge with a simple canal system in which openings in the body wall lead directly to the spongocoel, as in Leucosolenia.

Asconoid. Pertaining to sponges with the ascon type of canal system.

Ascon Type Canal System. Refers to a simple type of canal system. It is common in sponge Leucosolenia whose diploblastic body wall is perforated at regular intervals and flagellate cells (choanocytes) are lining the entire spongocoel.

Ascophora (class Gymnolaemata, order Cheilostomata). Suborder of cheilostome bryozoans in which the tentacles are protruded and withdrawn by means of a hydrostatic organ called a 'compensation sac'. The operculum is hinged in such a way that it controls the flow of water into the interior. The suborder occurs from the Cretaceous to Recent.

Ascophyllum (family Fucaceae, order Fucales). Genus of brown seaweeds which resemble Fucus, but which can be distinguished by the absence of a midrib. The thallus is typically flattened and strap-like, and air bladders are usually present; it may reach several metres in length. Spores develop in yellowish receptacles, resembling sultanas, borne on short stalks arising from either side of the fronds. The only British species is A. nodosum.

Ascorbic Acid (vitamin C). A lactone of a sugar acid that is found in high concentrations in certain fruits and green vegetables. It acts as a cofactor in the hydroxylation of proline to hydroxyproline. However its function in this reaction can be replaced by other compounds and it is not clear why it is essential to the growth of some vertebrates.

Ascospore. One of usually eight haploid spores characteristically formed inside an ascus. The two nuclei of the dikaryotic ascus initially fuse and undergo meiosis to produce four haploid daughter nuclei. These then undergo one mitotic division resulting in eight haploid nuclei. The cytoplasm of the ascus is then isolated around each nucleus to from the ascospores.

Ascostroma. A stroma in which ascus-containing chambers or locules develop.

Ascus. A specialised cell in fungi of the Ascomycotina in which ascospores develop. If it is cylindrical, as is usually the case, then the ascospores are violently discharged. However, in some ascomycetes (*e.g.* the Endomycetales and Eurotiales) the ascus is a globular and in these fungi the ascospores are released passively. The tip of the ascus may possess a cap, the operculum, or it may simply have a terminal pore. This characteristic is important in separating the operculate Pezizales from the inoperculate Helotiales. In most ascomycetes the ascus wall is a single layer (unituricate) but in the Loculoascomycetes it consists of two layers (bitunicate).

Asepalous. Lacking sepals.

Aseptate. Without septa.

Aseptic. Free of microorganisms, especially those causing decay, putrefaction, or disease.

Asexual. Lacking sex; not involving sex.

Asexual Reproduction. Reproduction in which new individuals are produced from a single parent without the formation of gametes. It occurs chiefly in lower animals, micro-organisms, and plants. In lower animals the chief methods are fission (*e.g.* in protozoans), fragmentation (*e.g.* in some aquatic annelid worms), and budding (*e.g.* in coelenterates). The principal methods of asexual reproduction in plants are by vegetative propagation (*e.g.* bulbs, corms, tubers) and by the formation of sexual spores. Spore formation occurs in mosses, ferns, and other plants showing alternation of generations, as a dormant stage between sporophyte and gametophyte, and in some algae and fungi, to produce replicas of the plant. Compare sexual reproduction.

Ash. A group of large trees found in Europe, Asia and North America. The common ash of Europe and south-western Asia is a majestic tree reaching a height of 40 metres. It has large leaves made up of 9—15 leaflets. It produces tough, elastic timber.

Ash, Common. A tall deciduous tree, up to 40 metres high, belonging to the olive family. It has pale grey bark, and is easily recognizable in winter by its black buds. The

Common ash

small purple male and female flowers usually grow on separate trees and open before the leaves. The winged seeds hang in bunches and are called keys. The mountain ash is an unrelated species of tree, also known as Rowan.

Asity. A fruit-eating bird of Madagascar, as the velvet asity (Philepitta castanea).

Asp. A poisonous snake closely related to the adder and distributed over the warmer parts of Europe. Its colour ranges from greyish-brown to orange or red, and it rarely exceeds 60 cm in length. It can be distinguished from the adder by its turned-up snout and by its yellow eyes. The adder's eyes are coppery-red.

Asparagine. An uncharged polar amino acid with the formula $NH_2COCH_2CH(NH_2)COOH$ (illustration at amino (acid). Asparagine is formed by the ATP-assisted addition of ammonia to aspartic acid. The reverse of this reaction (but without ATP formation) is the route of asparagine breakdown.

Because ammonia is extremely toxic to living cells, it is used to synthesize glutamine and asparagine, in which form it can be stored for later use in amino acid synthesis.

Asparagus (family Liliaceae). Genus of small shrubs or perennial herbs with creeping underground stems, and green, needle-like branchlets replacing the leaves, which are reduced to papery scales. A. officinalis is cultivated for its edible young shoots, which are then forced and blanched. A. plumosus produces attractive, feathery foliage used in flower bouquests. There are about 300 species found throughout the Old World.

Aspartic Acid. A acidic amino acid with the formula $HOOCCH_2CH(NH_2)COOH$ (illustration at amino acid). It is formed by a transmination, in which the amino group of glutamic acid is transferred to oxaloacetic acid so forming aspartic acid. Breakdown is by the reverse reaction, followed by further oxidation in the TCA cycle.

Aspartate is an important precursor of nitrogenous compounds. The amino acids leucine, isoleucine, threonine, methionine, and lysine all derive from aspartic acid, and it is necessary for synthesis of purines, pyrimidines, and porphyrins. Aspartate also acts as an amino-group donor in several reactions.

Aspect

1. The general appearance or view with respect to (*a*) an object or individual; (*b*) a species or population, especially its seasonal appearance.
2. An ecological season, as hiemal, vernal, aestival autumnal,

Aspen. A type of poplar, with rounded leaves held on long stalks, which quiver even on apparently still days. The male and female catkins grow on separate trees, as in all Poplars.

Aspergillales. Eurotiales.

Asphyxia. Suffocation; coma or apparent death resulting from lack of oxygen.

Aspidistra. A plant with long glossy leaves, native to China and Japan. It was an extremely popular indoor plant in dark Victorian houses because it tolerates poor light.

Aspidium. Male fern, the rhizome and stipes of Dryopteris filix-mas, the source of oleoresin of aspidium, a drug widely used for the expulsion of worms, especially tapeworms.

Aspidogastrea (Aspidogastraea; phylum Platyhelminthes). Class, or in some classifications as order of Trematoda, of parasitic, platyhelminth worms in which the gut, pharynx, intestine, and ventral sucker are present and the anterier sucker absent. Mature individuals also lack cilia. All are endoparasitic, and development may involve one or two hosts.

Aspirator. An apparatus such as a suction pump in which a negative pressure is created, used for the withdrawal of liquids or gases from a cavity.

Aspleniaceae. Family of ferns, distinguished among the advanced (leptosporangiate) ferns by its oblong or linear superficial sori, borne along one or both sides of the veins; the indusium is often much elongated, sometimes double, or absent. There are nearly 1C0 genera, of cosmopolitan distribution.

Aspredinidae (banjo catfish; class Osteichthyes, subclass Actinopterygii, order Siluriformes). Family of mainly small freshwater catfish whose common name is derived from the peculiar shape of the flat, triangular head region, and the long tail. In some species the female takes care of the fertilized eggs by carrying them along in a modified region of the skin of the belly. There are about 50 species, found in S. America.

Ass. The ass is a small kind of horse. Two species are known—one in North Africa and one in Central Asia. A race of the African species has been domesticated for a long time and we know it as the donkey. There may well be no truly wild asses left in Africa today, but the Asian species survives in several countries.

Assassin Bug. There are more than 3000 kinds of assassin bug, so called because of the violent way they seize and

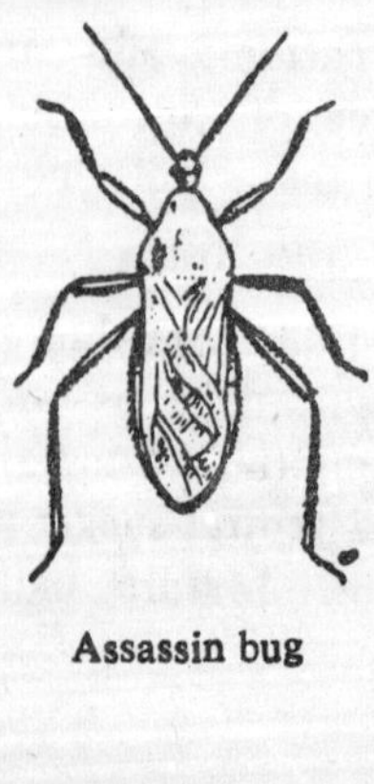

Assassin bug

kill their insect prey. Assassin bugs have a powerful curved beak which is thrust into the prey and which then sucks up their juices. Many assassin bugs mimic their prey and mix with them freely without causing alarm. Most species live in the tropics.

Assassin Fly. Robber-Fly.

Assimilation. The process of converting simple food materials into complex tissues of an organism.

Association. An ecological unit in which two or more species occur in closer proximity to one another than would be expected on the basis of change. Early plant ecologists recognized associations of fixed composition on the basis of the dominant species present (*e.g.* a coniferous forest association). Association now tend to be detected by using more objective statistical sampling methods.

Association Analysis. Hierarchical method of classification which is divisive and monothetic. Association analysis uses χ^2 as a measure of association between pairs of species (attributes) found at a range of sample sites (individuals). The species or attribute with the highest overall sum of χ^2 valuses (*i.e.* the strongest links) with all other species is selected as the basis for subdivision into two groups of sites or of individuals, either having or lacking that attribute. The process is then repeated for each new group until no further subdivision is required.

Association Measure. Any measure of the link between two variables as shown by quantitative or qualitative data describing their characteristics. Links between continuous variables—*e.g.* mass and height, or crop yield and rainfall—may be assessed statistically using correlation and rergession methods. Links between qualitative characteristics—*e.g.*, eye colour and hair colour, or aspect and slope form—may be assessed statistically using contingency tables and tests based on χ^2 or an appropriate non-parametric test. Sometimes continuous variables can be more conveniently tested in qualitative from, *e.g.*, height as short, medium, or tall, especially when the link with some other qualitative variable is wanted *e.g.*, aspect and tree height.

Association Neuron. A neuron especially on which connects various regions or centers of the central nervous system.

Associes. A seral plant community.

Assortive Mating. Breeding that is nonrandom. If similar phenotypes breed together it is positive assortive mating and leads to inbreeding. If dissimilar phenotypes breed together it is negative assortive mating and leads to outbreeding. The pollinating mechanism often has a major influence on the breeding patterns found. Thus in insect-pollinated plants, the pollinator may preferentially take pollen to flowers of a certain colour or with certain markings. In wind-pollinated plants crossing is more likely to be random but there is a tendency for plants of the same height to cross.

Assumentum. One of the valves of a silique.

Assurgent. Rising; ascending.

Astacus. A genus of crayfish. Species are widely found in fresh water and are omnivorous.

Asticcacaulis. Genus of Gram-negative bacteria which is not assigned is any taxonomic family. It is similar to Caulobacter in its life cycle, but differs in that the stalk occurs eccentrically at one pole or on one side of the cell, and does not function in attachment. Cells adhere to a substrate by adhesive material near one end of the cell. It is found in habitats similar to those of Caulobacter.

Astigmatism. Refers to a lens defect in which when rays in one plane are in focus those in an other plane are not. The eye may suffer from astigmatism, when the cornea is not spherical. It is corrected by using an anastigmatic lens, which is having different radii of curvature in the vertical and horizontal planes.

Astasia. Inability to walk or stand due to motor incoordination.

Astaxanthin A xanthophyll pigment (a hematochrome) present in euglenoids and chlorophytes.

[1]**Aster**. *bot*. A fall-blooming flower of the genus Aster.

[2]**Aster**. *cytol*. A star-shaped structure consisting of a centriole with radiating fibers appearing during mitosis at the two ends of the spindle.

Asteraceae. Compositae.

Asteriadina (starfish; subclass Asteroidea, order Forcipulatida). Suborder of starfish (asteroids) in which numerous dorsal (aboral) ossicles form crossed pedicellariae, consisting of a basal ossicle which bears a pincer-like pair of small spines. They first appeared in the Lower Jurassic, and the suborder includes many extant, predatory species.

Asteridae (Sympetalae). A subclass of the dicotyledons containing mostly herbaceous plants having flowers with fused petals. They have few stamens and usually only two carpels and the seeds are surrounded by only one integument (unitegmic). The following seven orders are generally recognized; Gentianales (Contortae) including the Loganiaceae or Buddlejaceae (*e.g.*, buddleias), Gentianaceae, Apocynaceae (*e.g.*, oleanders), and Asclepiadaceae (*e.g.* milkweeds); Polemoniales, including the Convolvulaceae (*e.g.*, bindweeds) Polemoniaceae (*e.g.*, phlox), and Boraginaceae (*e.g.*, forget-me-nots); Lamiales, including the Verbenaceae and Labiatae; Scro-phulariales, including the Scrophulariaceae (foxglove family), Gesneriaceae (*e.g.*, gloxinias), and Acanthaceae; Campanulales (Campanales), including the Campanulacease (bellflower family) and Lobeliaceae; Dipsacales, including the Caprifoliaceae (*e.g.*, honeysuckles; and Asterales, comprising the Compositae. Polemoniales members of the Lamiales, and Scrophulariales are sometimes grouped in the Tubiflorae.

Three further orders are sometimes recognized; Plantaginales (*e.g.* plantains) often included in the Scrophulariales; Rubiales, often included in the Gentianales; and calycerales, often included in the Dipsacales.

The affinities of the important family Solánaceae are uncertain, some placing it in the Polemoniales and others in the Scrophulariales.

Asteroidea. The class of the Echinodermata which is including the starfish (*e.g.* Asterias). They are usually found just below the low-tide mark. A starfish typically has five arms radiating from a central disc, which contains the main body organs and has the mouth on its ventral surface. Chalky plates in the skin form a skeletal test. Suckered tube feet on the underside of the arms take part in locomotion and holding prey.

Astrapotheria (cohort Ferungulata, superorder Protungulata). Extinct order of S. American ungulates which appeared during the Eocene and flourished in the absence of herbivore competition or carnivore predation during the Oligocene and Miocene. The skull was short, the upper incisors were lost and the premaxillae much reduced, the lower incisors were present, the upper and lower canines were large and persistently-growing, and the rearmost molars were large. A slit-like nasal opening on top of the skull suggests that astrapotherids—*e.g.*, the Miocene Astrapotherium, which was the size of a rhinoceros—had elephant-like probosces. In contrast to the heavily built front legs, head, neck and anterior part of the body, the spine, hindquarters, and hind legs were feeble. The feet were small, perhaps resting on pads, but unlike these of any other group. Possibly, therefore, the animals were amphibious.

Astroblepidae (class Osteichthyes subclass. Actinopterygii, order Siluriformes). Family of freshwater fish with naked skin small eyes and ventrally positioned mouth. There are about 35 species, occurring in S. America.

Astrocoeniina (order Madreporaria). Suborder of colonial corals with small corallities, which appeared first in the Middle Triassic. They are characterized by the possession of septa formed from relatively few (not more than eight) trabeculae, and usually they have no more than two cycles of tentacles arranged in a single ring around the mouth, which leads to a smooth stomodeum. The suborder contains five families, with 46 genera.

Astrosclereid (asterosclereid, star sclereid). A relatively short sclerenchyma cell (sclereid), differing from a brachy-sclereid by its often conspicuously branched shape. Astrosclereids are usually present singly or in small groups and are often found in the mesophyll of leaves, where they act as a strengthening agent.

Asteroid. An echinoderm of the class Asteroidea which includes the sea stars (starfishes); asteroidean.

Asthenia. Muscle weakness; loss of strength.

Astigmatic. Lacking stigmas.

Astigmatism. A lens defect in which when rays in one plane are in focus those in another plane are not. The eye can suffer from astigmatism, usually when the cornea is not spherical. It is corrected by using an anastigmatic lens, which has different radii of curvature in the vertical and horizontal planes.

Astomatous

1. *zool. a.* Lacking a mouth or stoma. *b.* Lacking a cytostome, said of holotrich ciliates.
2. *bot.* Lacking stomata.

Astragulus. The tarus, a tarsal bone.

Astrocyte A type of neuroglial cell with many branching cytoplasmic processes.

Asymmetrical Lacking in symmetry; not capable of being divided into equal or corresponding parts.

Asynchronous Muscle. A type of muscle which is found in thorax of some insects. Undergoes contraction without any one to one relation to motor impulses (also called fibrillar muscle).

Atactostele. The term used for describing the distribution of vascular tissue in those angiosperms in which the vascular bundles have been scattered in an apparently random fashion in the ground tissue. Each vascular bundle has been surrounded by a pericycle and endodermis. This arrangement is typical of the stem structure of monocotyledons but is also found in dicotyledons in which a complete network of interconnected bundles arises as the stem thickens with age.

Atavism. The recurrence of traits possessed by a remote ancestor, usually the result of a recombination of genes.

Ataxia. Refers to the incoordination due to errors in rate, range, force and direction of movement; occurred due to leisions in cerebellum. There occurs no abnormality at rest.

Ateliosis or **Ateleiosis.** Dwarfism with retention of infantile characteristics.

Atelostomata (class Echinoidea). Superorder of irregular echinoids characterized by the lack of compound ambulacral plates, lantern, girdle, or gill slits. The apical system and peristome are rarely opposite, and the interambulacra are wider than the ambulacra on the lower surface. The superorder includes the orders Disasteroida, Holasteroida, and Spatangoida.

Athecate. Lacking a theca.

Atherinidae (silversides, sand smelts; class Osteichthyes, subclass Actinopterygii, order Atheriniformes). Family of rather small, slender fish, with relatively large eyes, terminal mouth, and two widely separated dorsal fins. Most species occur in schools in inshore seas, with some found in estuaries or freshwater systems. There are more than 150 species. Perhaps the best known species is Leuresthes tenuis (California grunion), famous for its communal spawning on sandy shores at high tide at night.

Atheroselerosis. A form of arteriosclerosis in which localized deposits of lipids form beneath the intima of an artery.

Athyrium (family Aspleniaceae). Genus of ferns with large, soft rhizome scales, and 1—3 pinnate, soft, hairless fronds; sori are borne on receptacles on a side branch of a vein; indusia are hooked, or crescent-shaped, and toothed, or rarely absent. There are about 180 species. They are cosmopolitan, but most occur in temperate F Asia.

Atheriniformes (class Osteichthyes, subclass Actinopterygii). Order of bony fish, including fairly large marine individuals, *e.g.* the flying fish and the sauries, as well as many small freshwater species, *e.g.* killifish, guppy, and mosquitofish. Members of this order have also been placed in other orders, including the Beloniformes, and the Cyprinodontiformes.

Atlas. The first cervical vertebra, a ringlike bone that joins the skull to the vertebral column in terrestrial vertebrates. In advanced vertebrates articulation between the skull and atlas permits nodding movements of the head.

Atlas Moth. One of the largest moths in the world, with a wing span up to 25 cm, the atlas moth is found in India and other parts of South-East Asia. It is mainly brown, but there is a transparent spot on each wing. The atlas moth belongs to the same family (Saturniidae) as the European emperor moth and the Indian tussore silk moths, but it does no provide useful silk.

Atoke. In certain marine polychaetes, the anterior nonsexual form as distinguished from the posterior sexual form or epitoke, as in the palolo worm.

Atokous. Lacking offspring.

Atoll. Refers to the horse-shoe-shaped or ring-shaped island of a belt of coral reef which is surrounding a central lagoon.

Atmometer. Potometer.

ATP (adenosine triphosphate). A nucleotide that is of fundamental importance as a carrier of chemical energy in all living organisms. It consists of adenine linked to D-ribose (*i.e.*, adenosine); the D-ribose component bears three phosphate groups, linearly linked together by covalent bonds. These bonds can undergo hydrolysis to yield either a molecule of ADP (adenosine diphosphate) and inorganic phosphate or a molecule of AMP (adenosine monophosphate) and pyrophosphate. Both these reactions yield a large amount of energy (about 30.6 kJ mol^{-1}) that is used to bring about such biological processes as muscle contraction, the active transport of ions and molecules across cell membranes, and the synthesis of biomolecules. The reactions bringing about these processes often involve the enzyme-catalysed transfer of the phosphate group to intermediate substrates. Most ATP-mediated reactions require Mg^{2+} ions as cofactors.

ATP is regenerated by the rephosphorylation of AMP and ADP using the chemical energy obtained from the

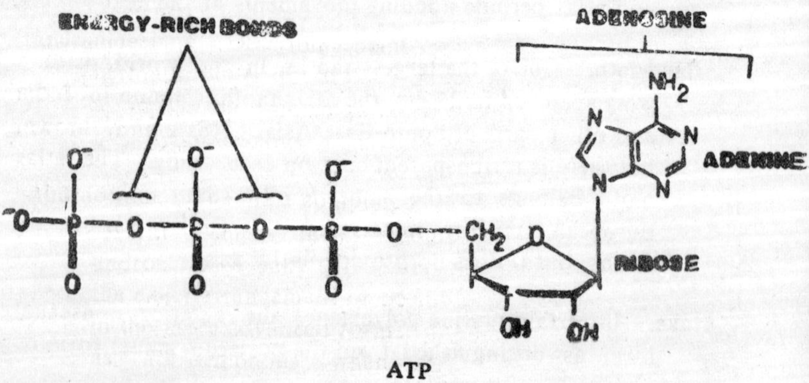

ATP

oxidation of food. This takes place during glycolysis and the Krebs cycle but, most significantly, is also a result of the reduction-oxidation reactions of the electron transport chain, which ultimately reduces molecular oxygen to water (oxidative phosphorylation).

Atresia. Imperforation or closure of a normal opening, canal, or cavity.

Atrial. Of or pertaining to an atrium.

Atrial Cavity. Used for describing a cavity in protochordates and some caudate larvae (*e.g.*, tadpole) through which water passes after moving out of gill slits.

Atrichornithidae (scrub-birds; order Passeriformes). Family of small, brown birds, with short, round wings, and long, graduated tails. They have powerful voices, especially Atrichornis clamosus (noisy scrub-bird). They are insectivorous, strongly territorial, and are found only in the dense undergrowth of two isolated areas of coastal Australia. There is one genus, containing two species.

Atrichous

1. Without spines, with reference to nematocysts.
2. Without flagella, with reference to bacteria.

Atrichous Isorhizas. Also known as steroline glutinants; oval in shape having small thread but bearing no spines. Takes part in locomotion of certain animals like Hydra by adhering with the substratum through hypostome.

Atriopore. Used for the external opening from atrial cavity in Amphioxus and related animal forms.

Atrioventricular Node (AV-Node). Refers to one of the conducting systems of heart situated in the right posterior portion of interauricular septum. There are three bundles of fibres which have Purkinjie fibres and found to conduct impulses from SA-node to this node (anterior) internodal tract of Bachman, middle internodal of Wenokebach and posterior internodal tract of Thorel). AV-node is continuous with bundle of His.

Atriplex (family Chenopodiaceae, tribe Cyclobeae). Genus of herbs and small shrubs which are generally halophytic. The stems are often straited with red or white, and plant is often mealy. The leaves are alternate, toothed or lobed, linear or triangular, and without stipules. The flowers are inconspicuous, unisexual, and held in small cymes. The perianth of the male flower is composed of 2—5 segments, usually green in colour, and there is a similar number of stamens. The female flower is without any perianth, but is enclosed by two persistent bracteoles, which can be entire or lobed. The superior ovary is composed of three fused carpels, and a single locule with one ovule. The fruit is an achene. There are about 100 species, found throughout the world, often as components of open communities or grassland, on soils rich in organic salts, *e.g.*, in coastal areas.

Atriopore. The opening of an atrial cavity to the outside.

Atrioventricular. Pertaining to the atria and ventricles of the heart. Abbreviated AV, A-V.

Atrium

1. (or auricle). A chamber of the heart that receives blood from the veins and forces it by powerful muscular contraction into the ventricle(s). Fish have a single atrium but all other vertebrates have two.

2. Any of various cavities or chambers in animals, such as the chamber surrounding the gill slits of Amphioxus and other invertebrate chordates.

Atrocha. A ciliated larva of certain polychaete worms.

Atrophy. The reduction in size of an organ or structure resulting from lack of nourishment or reduced functional activity.

Atropine. An alkaloid obtained from Atropa belladonna and other plants of the family Solanaceae. It is a parasympathetic blocking agent and used as a respiratory stimulant, muscle relaxant, and mydriatic (dilator of pupils).

Atropous. Designating an ovule which is not inverted with micropyle or nucellar apex directed away from the funiculus; orthotropous. Compare anatropous.

Attenuate

1. To make thin, dilute, or weaken.
2. *bacteriol.* To reduce the virulence of, with reference to pathogenic organisms.

[a]Attenuate. *bot.* With a long, gradual taper; accuminate.

Attenuation

1. (in medicine). A process of reducing the disease-producing ability of a microorganism. It can be achieved by chemical treatment, heating, drying, irradiation, by growing the organism under adverse conditions, or by serial passage through another organism. Attenuated bacteria or viruses are used for some vaccines.
2. (in mycology). The conversion by yeasts of carbohydrates to alcohol; as in brewing and wine and spirit production.

Attid. A jumping spider of the family Attidae.

Attini (order Hymenoptera, suborder Apocrita, family Formicidae, subfamily Myrmicinae). Tribe of fungus-feeding ants in which foliage cut by foraging workers provides a substrate for the growth of a fungas, localized in fungus gardens within the soil nest. Fungal gongylidia (hyphae with swollen tips) are harvested by workers and fed to larvae. Each mature colony of Atta sexdens may have three million workers, produced by a single queen whose life-span may be eight years or more. There are 200 species, restricted to the New World, where most occur in the neotropics.

Attitude. The position or posture assumed by an organism.

Atto-. Symbol *a*. A prefix used in the metric system to denote 10^{-18}. For example, 10^{-18} second=1 attosecond (as).

Atypical. Irregular; not according to type.

Atypidae (purse-web spiders; order Araneae, suborder Mygalomorphae). Family of spiders, all less than 3 cm long, which have three tarsal claws and six spinnerets. The species are robust, with massive chelicerae, and two pairs of book lungs. New World species construct open silk tubes on the sides of trees; European species construct the tubes on the ground. In all cases, prey walking over the tube is impaled on the fangs and dragged through the purse web. The three species of Atypus are the only Mygalomorphae found in N. Europe. Atypus is also found in the eastern VSA, Burma, Java, and Japan. The few species in this family may live for up to nine years.

Atypus Affinis (purse-web spider; suborder Mygalomorphae, family Atypidae). Species of purse-web spider which has a brown or greenish-brown. glossy carapace, and a dark reddish-brown abdomen that is marked with small grey spots. Females can attain lengths of 15 mm, and males 9 mm. Tube webs are usually made under cover of low vegetation in sandy or chalky soil. The species has been recorded throughout the UK, but is local in northern areas.

Aubergine (or egg-plant). A large plant producing purple egg-shaped fruits. It is a member of the potato family and a native of southern Asia, but the aubergine is grown as a vegetable in other warm places, and in greenhouses.

Audibility. The limits of audibility of the human ear are between about 20 hertz (a low rumble) and 20 000 hertz (a shrill whistle). With increased age the upper limit falls quite considerably.

Audiometer. An instrument that generates a sound of known frequency and intensity in order to measure an individual's hearing ability.

Auditory. Of or pertaining to hearing of the sense or sound, acoustic.

Auditory Capsule (Otic Capsule). The cartilaginous or bony part of the vertebrate skull that is enclosing the inner ear. It has a single periotic bone in adult mammals.

Auditory Meatus. One of two canals in the temporal bone, the external conveying sound waves to the middle ear, the internal transmitting the vestibulocochlear and facial nerves

Auditory Nerye (Vestibulocochlear Nerve, Acoustic Nerve, Cranial nerve VIII). The nerve supplying the inner ear. It is arising from the dorsal region of the medulla oblongata in the vertbrate brain. It consists of two main branches—the vestibular nerve, which is supplying the anterior region, and the cochlear nerve, which is supplying the posterior part (including the cochlea).

Auditory Organ. An organ capable of being stimulated by sound vibrations, especially an organ or hearing as *a.* in vertebrates, the ear; *b.* in invertebrates, tympanal organs and auditory hairs, as in insects.

Auditory Ossicles. The ear bones, malleus, incus, and stapes, located in the middle ear of mammals.

Auditory Tube. A tube connecting the middle ear with the nasopharynx. Also called Eustachian tube.

Auditory Vesicle. Otocyst.

Auerbach's Pleuns. Used for describing a network of small nerve fibres and ganglia between the two muscular coats of the intestine; able to cause local reflex arcs by which pressure of food in the intestine can bring about peristaltic movement.

Auger Shell. A slender, many-whorled gastropod of the family Terebridae, common in the Indo Pacific region. A common genus is Terebra.

Auk. A group of seabirds, including the littie auk, Puffin, Guillemot and Razorbill. They are black and white birds, with slim, stubby wings, just as at home in the sea as in the air. Auks nest on sea cliffs, and spend most of their time at sea, catching fish.

Auklet. One of several species of small auks of the genera Ptychoramphus and Cerorhinco of the North Pacific.

Aulopodidae (class Osteichthyes, subclass Actinopterygii, order Myctophiformes). Small family of marine, warm- to temperate-water fish. All species possess a rather large head, big jaws, a high dorsal fin, and a slender body. Aulopus purpurissatus (Sergeant Baker) is well known in Australia. It is a beautifully coloured species of up to 50 cm body length. There are about eight species.

Aujostomidae (flutemouth, trumpetfish; class Osteichthyes, subclass Actinopterygii, order Syngnathiformes). Family of marine tropical fish characterized by a very elongate body, a long compressed snout, a series of isolated dorsal spines, and a rounded tail-fin. Found mainly in reefs or inshore rocky areas, they can be encountered swimming in an almost vertical position, hidden by seagrass leaves or coral branches. There are about four species.

Auricle

1. *zool.* An ear or earlike structure, as *a.* the pinna of the external ear; *b.* a receiving chamber of the heart in various invertebrates; *c.* the auricular appendage, a protruding portion of the atrium in the heart of mammals; *d.* a blunt sensory projection at the anterior end of turbellarians; *e.* a ciliated projection

near the mouth in ctenophores and rotifers; *f.* a lip or plate forming a part of the pollen packer in honeybees.

2. *bot.* An earlike lobe at the base of various leaves and petals.

[1]**Auricular**

1. Of or pertaining to the ear or the sense of hearing.
2. Shaped like an ear.

[2]**Auricular**. *ornith.* One of the feathers which covers the external auditory opening.

Auricular Feathers. Refers to a tuft of large feathers which are surrounding the ear of an owl or similar bird and giving the appearance of a pinna.

Auricularia (order Auriculariales). Genus of fungi in which the fruit bodies are gelatinous but firm, and may be cup-shaped, ear-shaped, discoid, etc. The hymenium is borne on one side of the fruit body. A. auricula (also known as Hirneola auriculajudae) is the 'jew's ear' fungus which is a common parasite on elder (Sambucus) and other trees. A. mesenterica has a resupinate fruit body with a velvety surface and a jutting, shelf-like upper margin. It is common throughout the year on hardwood stumps and logs. There are 10 speeies.

Auricularia Larva. Used for describing the typical larval form of a holothuroid. It has a band of cilia arranged in a complex pattern down each side and folded back at each end where the body has been extended to form a pre-oral and a post-anal lobe.

Auriculate. Possessing an auricle or auricles.

Auriform. Shaped like an ear.

Aurochs. The European bison or wisent (Bison bonasus) now extinct in the wild state. It has been considered to be the ancestor of modern breeds of cattle.

Aurones. One of the two groups of anthochlor flower pigments. They are probably formed by oxidation of chalcones. Aurones are found in many members of the Compositae, especially the genus Coreopsis.

Aurophore. A constricted portion of the float in certain siphonophores.

Auscultation. The art of listening to sounds produced within the body, as heart and respiratory sounds.

Australopithecines. Literally 'southern ape men', these early protohominids of the Pliocence and early Pleistocene (living 3—1.5 million years ago) are known only from Africa. Though placed formerly in a genus, Australopithecus, the 'robust australopithecines' are often referred nowadays to a genus Paranthropus, while the 'gracile australopithecines, are called Homo africanus.

Australopithecus. A genus of fossil primates that lived 1.3—4 million years ago. They walked erect and had teeth resembling those of modern man, but the brain capacity was less than half that of a human. Various finds have been made, chiefly in East and South Africa (hence the name, which means 'southern ape'). Australopithecus and related genera are known as australopithecines.

Austrilina (family Urticaceae). Genus of annual and perennial herbs which are often creeping. The leaves are alternate or opposite, with petioles and small stipules. The unisexual flowers are held in axillary clusters. They are small and green. The male flowers are single, or in groups of up to five on a common stalk The perianth is two-lobed and there is a single stamen. The female flowers are usually solitary or in a small cluster with a tubular, slightly toothed perianth, which persists until fruiting, along with the stigma. There are about five species, found in Australia, New Zealand, and S. Africa.

Autacoid. Term formerly applied to a hormone which stimulated the target organ.

Autapomorphy. Any derived character state (apomorphy) only possessed by members of one particular taxon. Autapomorphies distinguish a taxon from other related taxa. The number of autapomorphies possessed by one taxon as compared to another taxon assumed to be derived from the same common ancestor, provides a measure of anagenesis.

Autecology. The study of a single species and its relationship with the environment. It involves the investigation of the life cycle of the organism, recording the effects of the nonliving and living factors in the environment at each stage. Quantitative records are kept, particularly those relating to variations in numbers within and between populations. Compare synecology.

Author. In taxonomy, the person who published the first valid name of a taxon. The name of the author, or a recognized abbreviation of it, should follow the name of the taxon for it to be complete. Many plant taxa are followed by L. (for Linnaeus). If the rank of a taxon at the genus level or below is changed but the name or epithet retained then the name of the original author is also retained in brackets before the name of the person who made the change. For example, Medicago polymorpha var. orbicularis L. was raised to the species level by Allioni and is thus named Medicago orbicularis (L.) All. A similar procedure is followed if a taxon at the subgeneric level is transferred to another taxon but retains its epithet.

If a name proposed by one person but not published by him is published by another and ascribed to the first author then the taxon is followed by the name of the first author followed by ex (described by) and then the name of the publishing author. For example, the Corsican speedwell is Veronica repens Clarion ex DC. (DC. for De Candolle). The citation of authors in this way helps other workers find the original descriptions and type specimens.

Autocarp. A fruit developing from self-fertilization.

Autocatalysis. A process in which the speed of a chemical reaction is accelerated by the presence of one or more products resulting from the reation.

Autochthonous. Local, indigenous, native.

Autoclave. An apparatus within which very high temperatures are produced by steam under pressure. It is used to sterilize laboratory equipment and is essentially a large pressure cooker.

Autodiploid. A gamete containing a diploid rather than haploid number of chromosomes. Such gametes may arise either through faulty meiosis or from tetraploid tissue. They may fuse with another such gamete to give a tetraploid zygote or may develop parthenogenetically. The term is also applied to diploid plants that have been obtained by doubling the chromosomes of a haploid plant. Such plants are completely homozygous.

Autoecious. Requiring but a single host to complete its life cycle, with reference to rust fungi.

Autogamy

1. A type of reproduction that occurs in single isolated individuals of ciliate protozoan animals of the genus Paramecium. The nucleus divides into two genetically identical haploid nuclei, which then fuse to form a diploid zygote. The onset of autogamy is associated with changing environmental conditions and may be necessary to maintain cell vitality.
2. Self-fertilization in plants.

Autogenous. Arising from within self generated; endogenous.

Autograft. A type of graft which involves transplantation of tissue or an organ from one part of an individual to another part of the same individual.

Autograph. Tissue removed from one part of the body and transplanted to another region of the same organism.

Autoimmunity. A disorder of the body's defence mechanism in which an immune response is elicited against its own tissues. Rheumatoid arthritis is basically an autoimmune disease.

Autologous. Obtained from or derived from an organism in contrast to homologous (from another of the same species) or heterologous (from one of a different species).

Autolysis. The process of self-digestion undergone by organelles or cells when their useful life is completed. It is effected by enzymes, primarily hydrolytic in character, that are produced by the cytoplasm. The process is believed to be under the control of growth substances. The products of the digestion are subsequently reabsorbed by the surrounding cells.

Automatism. Spontaneous activity of tissues, organs, or organisms.

Automimicry. Presence of a polymorphism for palatability to predators. In the unpalatable monarch butterfly, for example, it arises from the particular toxic food plants which are chosen by the ovipositing female. After hatching from the egg the larvae feed on the plant, taking in substances that are toxic to birds but not to the insects themselves. This renders them unpalatable to predators. The situation is used to advantage by palatable members of the species, which mimic the coloration of the unpalatable ones and so obtain protection. The polymorphism is frequency-dependent, and the unpalatable insects must be more abundant—and hence more frequently encountered by the bird predators—than the palatable insects.

Automixis. In certain protozoans, the fusion of two nuclei within a single cell; autogamy.

Autonomic. Self-regulating; functioning independently; spontaneous.

Autonomic Movement. A movement of a plant or plant part in response to a stimulus generated within the plant. Autonomic movements include the beating of cilia and flagella, protoplasmic streaming, circumnutation, and the movement of chromosomes during nuclear division. Compare paratonic movement.

Autonomic Nervous System (ANS). The part of the vertebrate peripheral nervous system that supplies stimulation via motor nerves to the smooth and cardiac muscles (the involuntary muscles) and to the glands of the body. It is divided into the parasympathetic and the sympathetic nervous systems, which tend to work antagonistically on the same organs. The activity of the ANS is controlled principally by the medulla oblongata and hypothalamus of the brain.

Autonomous. Capable of existing or operating independently; self-regulating.

Autophagy. A process utilized by certain cells to digest worn out or superfluous cell organelles. An autophagic vacuole forms around a portion of cytoplasm containing one or more unwanted organelles. Hydrolytic enzymes are secreted into the vacuole by the surrounding cytoplasm and digestion occurs followed by the subsequent reabsorption of the breakdown products.

Autophilous. Self-pollinated.

Autophyte. A plant capable of manufacturing its own food. Compare heterophyte.

Autoplastic. Of or pertaining to grafts transferred from one part of the body to another.

Autoploid. An autopolyploid.

Autopolyploid. A polyploid in which multiple genomes are identical as an autotetraploid.

Autopolyploidy. A form of polyploidy that results from the multiplication of chromosome sets from a single species. Autopolyploids may arise either through a failure in the mitotic process (syndiploidy), or in the meiotic process, resulting in the formation of diploid rather than haploid gametes. Autopolyploids often resemble their diploid parents, except that they may grow more slowly and flower later. Their cells are usually larger and the epidermis thicker. Autopolyploids with an uneven number of chromosome sets (autotriploids, autopentaploids) are often sterile because their gametes contain unbalanced number of chromosomes. Autotetraploids also show reduced fertility, the degree depending on whether any trivalents or univalents are formed at meiosis and also on the orientations taken up by the multivalents at metaphase. The patterns of inheritance, although they follow Mendelian laws, are very complex in autopolyploids because of the increased number of genotypes that may exist for any given locus with two alternative alleles. Thus for the two alleles A and a, a diploid can have three possible genotypes, AA, Aa, and aa, while an autotetraploid can have five, AAAA, AAAa, AAaa, Aaaa, and aaaa. In addition, while the frequencies of the gametes A and a can usually be predicted in diploids, the frequencies of the gametes AA, Aa, and aa of an autotetraploid are hard predict because segregation can occur at either the first or second meiotic division.

Compare allopolyploidy.

Autoradiograph. A radioautograph.

Autoradiography. An experimental technique in which a radioactive specimen is placed in contact with (or close to) a photographic plate, so as to produce a record of the distribution of radioactivity in the specimen. The film is darkened by the ionizing radiation from radioactive parts of the sample. Autoradiography is used to study the distribution of particular substances in living tissues and cells. A radioactive isotope of the substance is introduced into the organism or tissue, which is killed, sectioned, and examined after enough time has elapsed for the isotope to be incorporated into the substance.

Autorhytemicity. Refers to the generation of rhythmic activity without involving any extrinsic control *e g.*, activity in cilia and flagella and pace-makers (in the myogenic hearts). These structures beat even in the absence of nervous control.

Autorhythmic. Capable of spontaneous, rhythmical self-excitation.

Autosome. Either of any pair of chromosomes that do not play a principal role in sex determination. The term is used of plant chromosomes when discussing dioecious species that have cytologically identifiable sex chromosomes.

Autospore. A zoospore or aplanospore which has the same shape as the parent cell.

Autostylic. A method of jaw suspension by the quadrate, common in tetrapods other than mammals.

Autostylic Jaw Suspension. A type of jaw suspension. It is found in amphibians, reptiles, and some fish whereby the upper jaw gets attached directly to the skull.

Autotetraploid. Autopelyploid with four similar genomes. They occur either naturally by the spontaneous accidental doubling of a 2*n* genome to 4*n*, or artificially, through the use of colchicine. They are present in many commercially important crop plants because, as with other polyploids, they tend to be associated with increased size of the plant (through increased cell size, fruit size, stomata size, etc.).

Autotomy. The shedding by an animal of part of its body followed by the regeneration of the lost part. Autotomy is achieved by the contraction of muscles at specialized regions in the body. It serves as a protective mechanism if the animal is damaged or attacked (*e.g.* tail loss in certain reptiles) and is common as a method of asexual reproduction in polychaete worms, in which both new head and tail regions may be regenerated.

Autotroph. An organism that needs only simple inorganic compounds to grow. Thus carbon dioxide or carbonates serve as the carbon source and can be built up into complex organic molecules using light (phototroph) or chemical (chemotroph) energy. Simple inorganic nitrogen compounds are utilized in the synthesis of proteins. Most plants and some bacteria are autotrophic. Compare heterotroph.

Autotrophic

1. Self-nourishing, said of organisms which are capable of synthesizing food from inorganic compounds, as all chlorophyll-containing plants.

2 *microbiol.* Obtaining carbon from carbon dioxide or carbonates, as nitrifying bacteria of the soil.

Autotrophic Nutrition. Refers to the process of nutrition in which an organism is manufacturing is own food *e.g.*, nutrition in Euglena where animal is able to synthesise carbohydrates with the help of its chloroplast from CO_2 and water in presence of sunlight (as seen in the plants).

Autotrophism A type of nutrition in which organisms synthesize the organic materials they require from inorganic sources. Chief sources of carbon and nitrogen are carbon dioxide and nitrates, respectively. All green plants are autotrophic and use light as a source of energy for the synthesis, *i.e.* they are photoautotrophic (photosynthesis). Some bacteria are also photoautotrophic; others are chemoautotrophic, using energy derived from chemical processes. Compare heterotrophism.

Autozooid. A type of feeding polyp found in certain alcyonarians. Compare siphonozooid.

Auxanometer. An apparatus that is designed to measure the growth in length of plant organs. The organ is attached by a taut piece of cotton to the end of the shorter arm of a pivoted lever. Any growth results in the upward movement of the arm, which is magnified and recorded by the other longer arm of the lever as its tip moves across a calibrated scale, or as it traces the movement on a slowly rotating drum (kymograph).

Auxiliary. That which complements, aids, or helps.

Auxiliary Cell. A specialized cell in the gametophyte thallus of some red algae, *e.g.* Polysiphonia, situated adjacent to the carpogonium. Following fertilization of the carpogonium the zygote produces filaments that grow towards and fuse with the auxiliary cell. This process appears necessary for the production of carpospores.

Auxin. Any of a group of plant growth substances responsible for such processes as the promotion of growth by cell enlargement, the maintenance of apical dominance, and the initiation of root formation in cuttings. Naturally occurring auxins, such as indoleacetic acid (IAA), are synthesized in the shoot tips. Synthetic auxins, such as 2, 4-D and 2, 4, 5-T, are used as weedkillers.

Auxospore. The protoplast of a diatom before formation of a siliceous cell wall occurs. Auxospores are usually formed following a sequence of vegetative cell divisions in which the daughter cells become progressively smaller. The decrease in size results from the nature of diatom cell division whereby the two halves of the cell separate, each half containing a daughter cell. A new half wall is then formed, the edges of which fit within those of the old half wall in a manner resembling a Petri dish. Thus the diatoms decrease in size until a certain limit is reached, at which point auxospores are released. The naked auxospores are free to expand before forming a cell wall. Auxospores may also form after sexual fusion or by apogamy or parthenogenesis.

Auxotroph. A type of biochemical mutant that can grow normally if provided with a supplement in its diet to replace the substance it has lost the ability to make. For example, certain mutants of the mould Neurospora crassa need the amino acid arginine in the culture medium to grow normally (the unmutated mould can synthesize all the amino acids it requires from ammonia). By supplying the auxotrophic mould with various precursors of arginine it can be established which of the enzymes in the pathway to arginine is missing.

Avadavat. A small strikingly-coloured bird of South East Asia. The male has red and black plumage dotted with white spots.

Available Space Theory. The theory that the origin of new leaf primordia at a stem apex is governed physically by the amount of space between existing primordia. It is postulated that a new primordium only arises at a point where a certain minimum amount of space has become available. This is usually between the penultimately formed primordium and the one formed previous to that. The pattern that results from such restrictions is the same as would occur if inhibitors were released by older primordia.

Averrhoa (family Oxalidaceae). Genus of two species of small trees widely cultivated throughout the tropics for their edible, fleshy, many-seeded fruits. A. bilimbi (bilimbing) has acid fruits like small cucumbers; A. carambola (carambola) has elongate 5-angled fleshy berries, sweet or acid, and is also used for cleaning brassware.

Avena Curvature Test. A quantitative method used to correlate the degree of curvature obtained in a plant part, after the application of an external stimulus, with the amount of

and less commonly leaves. Cereals and grasses having ears or spikes covered in awns are termed awned or bearded. An individual organ bearing an awn is termed aristate. An awn may act to bury a fruit in the soil by uncoiling in damp conditions and is so doing pushing the fruit into the ground.

Axenic. Germ-free; free from foreign organisms.

Axenic Culture. A culture medium in which only one type of microorganism is growing. Such cultures are widely used in microbiology to determine the basic growth requirements or degree of inhibition by antibiotics or other chemicals of a particular species.

Axial. Describing cells derived from the fusiform initials, which are elongated parallel to the long axis of an organ. The axial system refers to all these cells collectively.

Axial Filament. Axial rod.

Axial Gland. A spongy mass forming part of the haemal system of echinoderms; axial organ.

Axial Rod. The central supporting structure of an axopodium; an axial filament.

Axial Skeleton. The skull, vertebral column, sternum, and ribs.

Axil. The angle between a branch of leaf and the stem it grows from. Axillary (or lateral) buds develop in the axil of a leaf. The presence of axillary buds distinguishes a leaf from a leaflet.

Axile. Of, or located in an axis, with reference to placentation in plants.

Axile Placentation. A form of placentation in which the placentae arise along the central axis of the ovary. It is seen in compound ovaries that are divided by septa into a number of locules, as in Hyacinthus.

Axilla

1. *anat.* The armpit or axillary space.
2. *ornith.* The corresponding region under the wing of a bird.

Axillar. Axillary.

[1]Axillary

1. *anat.* Of or pertaining to the axilla.
2. *bot.* Of or located in an axil.

[2]Axillary. *ornith.* An axillar, one of a group of feathers occupying the region between wing and body of a bird.

Axis. The second cervical vertebra, which articulates with the atlas (the first cervical vertebra, which articulates with the skull). The articulation between the axis and atlas in reptiles, birds, and mammals permits side-to-side movement of the head. The body of the axis is elongated to form a peg (the odontoid process), which extends into the ring of the atlas and acts as a pivot on which the atlas (and skull) can turn.

Axis Cylinder. Used for rod-like central portion of a medullated nerve fibre.

Axis Deer. The two species of axis deer live in India and South-East Asia. One species, known as the chital or spotted deer, is a small, dainty animal with a white-spotted red coat. The other axis deer is quite different. Its squat piglike appearance has given it the name hog deer. Both species of axis deer have slender antlers, up to one metre long in the chital, but only 30—40 cm in the hog deer.

Axolatl. The larva of the tiger salamander (Ambystoma tigrinum), which is capable of breeding while still retaining its larval form.

Axon. The long threadlike part of a nerve cell (neurone). It carries the nerve impulse (in the form of an action potential) away from the cell body of a neurone towards either an effector organ or the brain.

Axoneme. The central (9+2) core of microtubules which are found in cilia and flagella. It consists of nine pairs of outer microtubules surrounding two single central microtubules.

Axoplasm. The cytoplasm or neuroplasm of an axon.

Axopodium. A semipermanent type of pseudopodium which possesses a central, semirigid supporting rod, as in various heliozoans.

Axostyle. A central supporting filament or rod in certain protozoans, as in trichomonads.

Aye-aye. A nocturnal lemur (Daubentonia madagascariensis) of Madagascar.

Aysheaia Pedunculata. Described by Walcott, this species is one of a few referred to the arthropod order Protoonychophora. It possessed an elongate, worm-like body with a terminal mouth and frontal pupillae. The body was segmented and bore 10—43 pairs of walking legs. Although marine, Aysheaia is considered a likely ancestor for the extant onychophorans that exist in humid tropical forests.

Aythya (pochards; family Anatidae). Genus of diving ducks with the feet set well back, and a lobed flind toe. Most inhabit fresh water, but some feed in sea water. Most are omnivorous. They nest on the ground. There are 12 species, found mainly in the northern hemisphere where they are migratory, but also in Australasia and Madagascar.

Azalea, Wild. A colourful flowering plant related to the heather. It is a low creeping shrub which forms a mat of tiny pink flowers, and is commonly found high up on mountains and no Arctic heaths.

Azollaceae. Specialized family of free-floating, aquatic ferns of uncertain affinities. The branched stems bear roots, and two-lobed, overlapping leaves. Sori bear either microsporangia or megasporangia. The megasporangia are each provided with floats; microspores are numerous and grouped together into 'massulae' which bear barbed hairs (glochidia) that hook on to the megaspores. There is one genus, Azolla, with six species, of tropical and warm-temperate distribution in America and Asia. They are not native in Europe, but are naturalized there.

Azoic. Without life, especially with reference to geologic time antedating the origin of life on the earth.

Azonal Soil. Soil.

Azotobacter (family Azotobacteraceae). Genus of bacteria characterized by the production of differentiated resting cells called 'cysts'. Vegetative cells are ovoid in shape and can carry out nitrogen fixation. There are several species, found in soil and water.

Azotobacteraceae. Family of bacteria in which the cells are typically Gram-negative, and rod-shaped or variable in shape. They are chemo-organotrophic, and can grow only in the presence of air. Species are capable or converting atmospheric nitrogen into an organic form that can be incorporated into the cells (a process called nitrogen fixation). They are saprotrophic, and are found in soil, water, plant material, etc. There are several genera.

Azygote. An individual which is developed without involving fertilization, from haploid individual.

Azygospore. A zygospore which is formed parthenogenetically, and is a resting spore. The term has been sometimes confined to zygospores formed from gametes.

Azygous. Unpaired; existing singly.